# MONEY MACHINES

*As if proving Heidegger correct, what the recent financial crisis clearly demonstrates is the extent to which the techniques and technologies of global finance have remained transparent and virtually invisible. In this eye-opening book, Mark Coeckelbergh expertly exhibits and examines the influential but often unseen machines, machinery, and mechanisms of money that now regulate every aspect of contemporary life.*

David J. Gunkel, Northern Illinois University, USA

*Mark Coeckelbergh is recognized internationally for illuminating the manner in which information and communication technologies (ICTs) create new forms of "distancing" and in particular 'moral distancing'. This important book extends that analysis to underscore the hidden ways ICTs shape money and global finance, alter relationships, and undermine responsibility.*

Wendell Wallach, Yale University, USA

# Money Machines

## Electronic Financial Technologies, Distancing, and Responsibility in Global Finance

MARK COECKELBERGH
*De Montfort University, UK*

ASHGATE

Published by
Ashgate Publishing Limited
Wey Court East
Union Road
Farnham
Surrey, GU9 7PT
England

Ashgate Publishing Company
110 Cherry Street
Suite 3-1
Burlington, VT 05401-3818
USA

www.ashgate.com

**British Library Cataloguing in Publication Data**
A catalogue record for this book is available from the British Library

**The Library of Congress has cataloged the printed edition as follows:**
Coeckelbergh, Mark.
Money Machines: Electronic Financial Technologies, Distancing, and Responsibility in Global Finance / by Mark Coeckelbergh.
pages cm
Includes bibliographical references and index.
1. Finance – Information technology. 2. International finance. 3. Electronic funds transfers. 4. Money. I. Title.
HG173.C647 2015
332.1'78–dc23 2014042389

ISBN 9781472445087 (hbk)
ISBN 9781472445094 (ebk-PDF)
ISBN 9781472445100 (ebk-ePUB)



MIX
Paper from responsible sources
FSC® C013985
FSC www.fsc.org

Printed in the United Kingdom by Henry Ling Limited, at the Dorset Press, Dorchester, DT1 1HD

# Contents

*Acknowledgements* *vii*

1 Introduction: The question concerning financial technologies and distance 1

2 From clay tablets to computational finance: A brief history of financial technologies and distancing 17

3 The pure tool that distances: Simmel's phenomenology of money and its relation to philosophy of technology 33

4 Geography 1: Financial ICTs and the global space of flows 63

5 Bitcoin and the metaphysics of money 89

6 Money machines and moral distance: Financial ICTs, automation, and responsibility 107

7 Geography 2: Placing, materializing, humanizing, and personalizing global finance 123

8 Resistance and alternative financial technologies and practices 151

9 Conclusion 177

*References* *187*
*Index* *197*

# Acknowledgements

I wish to thank the editor from Ashgate, Neil Jordan, for supporting my plan to write this book. I am also grateful to John Donker, with whom I discussed high-frequency trading, and to the people in the financial sector I talked to. Special thanks go to Laura Fichtner, who helped me with the research for this book. Finally, I wish to thank my wife and children for their patience during the last stage of writing.

# Chapter 1

# Introduction: The question concerning financial technologies and distance

## 1.1. Background, aim and approach: The question concerning financial technologies and distance

Since the financial crisis of 2008, the financial sector has come under considerable criticism. Unprecedented attention has been paid to ethical issues in banking and finance. For example, bankers have been accused of being greedy if they accept high bonuses, the Occupy movement has blamed the financial sector for creating social and economic inequality and injustice, and in Europe there has been protest against financial cuts in public spending and against supporting failing banks with public money. Politicians and professionals in the financial sector have been blamed for practices that are seen as destructive for solidary, justice, and trust. They are asked to critically reflect on their practices and reform them.

However, as the dust settles, not only bankers and politicians have to think about what they are doing. It is easy to blame 'them' and ask 'them' to solve problems; but what can 'we' do? What can *I* do? Answering these questions requires knowledge about finance and its ethical and social aspects. But do we have that knowledge? The controversies have also revealed a blind spot in what most of us know about contemporary finance: while we have become increasingly vulnerable to the ebb and flow of global finance and are becoming more aware of this vulnerability, we know very little about it – let alone about how to tackle the problems.

To some extent, this 'we' even includes professionals working in the field. We live in a complex global world full of risk and uncertainty, with many things going on that are hard to grasp, oversee, or manage. Today citizens, but also journalists, policy makers, and many financial experts are increasingly confronted with ethical, social, and political problems raised by a range of financial systems, processes and activities that they neither fully understand nor control. This is a serious moral and democratic lacuna that needs to be urgently addressed. More research on the ethical, social, and political aspects of finance is badly needed if we, as individuals and as societies in a globalized world, are to take responsibility for what is going on and take appropriate action to make things better. We need to know what these financial changes mean for the lives of people and for societies, and we need to develop visions and policies about how to cope with the current problems and how to deal with future ones.

This requires, among other things, that researchers in the humanities and the social sciences attend to a topic that has received considerably less attention from commentators in the media and from politicians, and indeed from most academics: the important role *technology* is playing in finance, in particular contemporary information and communication technologies (ICTs). Usually this topic is seen as a 'technical' one that has little to do with ethics or politics. This is also how financial problems in general are often seen, and how ICT problems are usually seen. It is assumed that it is a matter that can safely be delegated to experts and that it is far removed from questions regarding values and politics. But this assumption is mistaken.

Consider the practice of high-frequency trading (HFT). HFT uses computer algorithms to trade securities at high speed. A significant part of that trade turns out to be delegated to *machines*. Is this ethically acceptable at all, and, if so, should it be regulated and how? Most people think only humans can be held responsible for what they do. A responsible practice therefore seems to require that agency remains human. Is responsible financial practice still possible if machines take over key tasks in trade and finance? There are many more, and broader, ethical and philosophical questions regarding the relation between contemporary ICT and finance. For example, what is the influence of internet and internet-based technologies on financial practices? What does global finance mean in the context of the information society, indeed 'information revolution'? How does electronic globalization shape the way we invest? What is the influence of electronic money on the way we think about goods and their value? Is it good to buy virtual goods? Is 'virtual theft' problematic? If the internet makes possible new kinds of monetary systems, which ones should be supported? What is the nature and ethical significance of money anyway? Is 'electronic' money real? Is money a means or has it become an end? Do we value money in itself? Or is it a *medium* and indeed a *technology*? Does money bring people together or does it tear them apart? Does electronic and globalized trade change our societies? What is the influence of social media on financial practices? What are the implications of a common currency for solidarity and justice? Are traders and investors using electronic technologies responsible for the consequences of their decisions on people in distant countries, who may remain invisible in the electronic world of finance, but who might be affected by, for instance, a rise in food price? Who is responsible when financial machines fail – when an error occurs – and when this results in a financial crash with global impact? Raising, analysing, and discussing these questions is vital for understanding how the ICTs of today shape global finance and for tackling the normative challenges this poses.

Thus, electronic ICTs play a key role in contemporary finance, and if we are to tackle the ethical challenges raised by global finance, we have to know more about these technologies and their relation to ethics in finance, to society, and to our lives. Academic research and reflection can and must contribute to meeting this urgent societal need.

Philosophers can substantially contribute to this aim by developing a hermeneutically and normatively powerful conceptual framework that helps us to better understand and evaluate financial technologies, in particular financial ICTs. However, in the field of philosophy of technology and computer ethics surprisingly little attention has been paid to *financial* technologies and their ethical and societal impact, and in ethics of finance the role of *technology* – in particular financial ICTs – in relation to ethics has also been largely neglected. In terms of approach and topic, the primary academic contribution of this book lies in taking up that challenge and in making that connection: it brings together on the one hand insights from philosophy of technology and social studies of finance, and on the other hand concerns from ethics of finance in order to reflect on the role of ICTs in finance and the impact these technologies have on our lives and on society.

In so far as this philosophical and interdisciplinary effort is an 'ethics', it is not ethics in the sense of a list of straightforward recommendations for reform or a 'Code of Ethics' professionals can use (e.g., financial experts or designers of ICT). Such normative tools are useful and philosophers can play a role in developing them, but in this book I first try to understand the deeper epistemic and normative influence of financial technologies: how do current financial technologies shape our relation to the world and to others? The book asks 'the question concerning technology', to use the title of Heidegger's famous essay (Heidegger 1977), but the focus is on a specific type of technology and, as I will explain below, on a very specific question.

Fortunately, some social scientists have already done very helpful work on understanding the social meaning and impact of financial technologies, and philosophers can benefit from that work. The book is therefore also set up as an interdisciplinary project that involves learning from social-scientific studies in sociology, anthropology, geography, social studies of finance, and social studies of science and technology. However, there is a crucial difference in approach. I will combine philosophical reflection with results from these studies in order to paint a picture that is both broader and more explicitly normative than most work in the social sciences: this book does not hesitate to *evaluate* the role of money and other technologies in the history, present, and near future of finance ('Will the machines take over?'), and bluntly engages with broad-stroked claims about developments in human society and culture by constructing and critically discussing a specific argument about financial technologies and distance – not only physical distance, but also social and moral distance. This is an argument that starts from the intuition that there is something deeply problematic about financial technologies and their impact on our relations to others and to the world, that these problems are not only due to the behaviour of people (e.g., bankers, traders) or to 'the system', but also, and crucially, have to do with the technologies used. It is an argument that learns from classic authors such as Simmel (1907) but also benefits from reflection in philosophy of technology about how technology changes human experience and society (e.g., McLuhan) and about engagement and detachment, for example, in Dreyfus and Borgmann. The notions of 'distancing' and 'moral distance' will be used to articulate and analyse this argument and to further inquire into the phenomenology of money and financial technologies. However, based on this analysis and in tune with contemporary ethics of technology I will also take a more constructive and optimistic approach and explore alternative financial technologies and practices.

Thus, this book's aim is to better understand and evaluate financial practices by focusing on the problem of the relation between financial technologies and various forms of distancing. By discussing problems of distancing raised by the use of contemporary information and communication technologies in financial practices and their social, political, and geographical context, it will contribute to a better understanding of the epistemic and moral influence of financial technologies and will help to define under which conditions more responsible financial practices are possible. It will thus contribute to efforts that aim to improve the quality of financial technologies, the quality of financial practices, and the quality of the lives of people and of the societies they live in.

This focus on the issue of distancing will not only benefit further reflection on financial technologies and ethics of finance. The book's broader academic aim lies also in contributing to thinking about the ethical and social implications of ICTs in general. The concepts of 'distancing' and 'moral distance' are not only a helpful way to conceptualize and discuss the hermeneutic and normative problems raised by contemporary financial technologies; they also help to think about what ICTs do to our relation to the world and to others. Let me already at this point explain why the notions of 'distancing' and in particular 'moral distance' can be useful conceptual tools in this area.

Technologies are often *tele*technologies; that is, they are intended to bridge physical distance. For example, a spear bridges the distance between the hunter and the animal. Cars, trains, and aeroplanes bridge distances between places. This observation is also and especially relevant in the case of ICTs and especially ICTs in 'the information age'. Consider the telephone and television, but also internet and internet-based technologies and media such as smartphones and social media. They are meant to connect people, to bridge the distance between people, and to bring information closer to the user. At the same time, however, there are all kinds of *unintended* effects of the technology. One worry is that these technologies do not *really* bring people closer to one another and that they instead *create all kinds of distances*, including social distance and what I propose to call 'moral distance': people may become more distant from one another and from the world, they may care less about others and about the environment, they may engage less with one another and with their social and natural environment. For example, one may worry that rockets and drones make it easier to kill at a distance, that telemedicine reduces human contact, that information technologies do not encourage us to engage with our social and natural environment, that social media alienate us from those that are near to us, and that the internet threatens social cohesion and real democratic participation. Thus, I am interested in the questions: what kind of knowledge and experience is created by contemporary ICTs, and what are the moral and social implications in terms of 'distance' and (dis)engagement?

Another kind of distance which has moral consequences is one between humans and their activities: contemporary ICTs mediate not only our perception and experience of others and of the world; they also mediate our actions. Electronic devices play an important role in our lives: we work with them, we live with them. They change what it is to work, what it is to travel, what it is to walk, and so on. They shape how we do things. This raises the question of who or what acts. Is it still we who use the internet, or does the internet use us? Is this still human action, or does the technology itself take on some kind of agency? Are we still driving cars, or, if cars are increasingly automated, is the car driving us? Who is responsible when something goes wrong? One might be worried that there is an increasing distance between us and our activities, and that this alienates us from what we do and makes it at least more difficult to exercise responsibility. Are these moral and social concerns justified? And if 'distance' is created in these ways, then does that distance increase with time? What kind of processes of distancing are there, and what is their moral and social significance? Discussing such questions may not only help us to directly address the moral questions, for example, the question regarding moral responsibility; it may also help us to better understand technology and the moral, social, and political significance of technology – including our contemporary ICTs.

Financial technologies in the context of global finance are an interesting case to investigate problems of distance, including the various more specific forms of distance and distancing illustrated above. On the one hand, contemporary financial technologies bridge distances: between financial centres and between traders, between investor (and their location) and the object of investment (and its location), between traders and their customers, and so on. The technologies also bring information about markets closer to the traders, and enable remote action in the sense that computer programs can take over trading decisions and actions. These ways of bridging make possible financial globalization. Thus, financial technologies function as a medium that bridges physical distance; contemporary financial ICTs are designed and intended to make possible global exchange of financial data and to automate financial actions. On the other hand, the medium is also the message (McLuhan 1964); technologies and media do more than they intend and have consequences for human experience, human action, and society. ICTs change how we live our lives and they change how we think, and this is also true for financial ICTs. They are not 'mere media' or 'mere means'; they change and have changed what it means to engage in financial action. The globalization of finance, made possible by electronic ICTs, for instance, raises moral and social questions. In particular, one may be concerned that this technological-geographical development in finance also creates 'moral' distance in the sense that it creates barriers for knowledge and action, which has consequences for moral responsibility. How can financial decisions have sufficient ethical quality if traders are alienated from other traders and their contexts, from the objects of their investment and its stakeholders, from the markets they influence and the other financial contexts their actions are related to (e.g., food market, housing market, currency exchange, credit market, etc.), from the decisions made by algorithms? What kind of relation do traders have to these stakeholders and indeed to their own trading activities, and what does this mean for the extent to which they can exercise responsibility? Do traders, who have a lot of financial information, know much about the distant moral and social consequences of their actions? Do these professionals, swamped by numbers and anonymous data, have too little *moral and social* information? Does the mediation by a chain of financial ICTs result in emotional distance, in a lack of care and concern for others which may be affected by the trade? Is there a growing distance between monetary value and the value of goods such as real estate and food? Do derivatives and other financial products, in so far as they are bets on future prices, create temporal distance, and what are the financial and moral implications of this? Does 'machine' trade alienate financial professionals from their activities? Do they still control them? Do they still *understand* them? Or are they only understood by those who designed them? Do these designers understand the moral and social implications of their algorithms? How can democratic politics function with regard to finance if there is a huge distance between the knowledge financial experts have and the knowledge others have? Is the 'financial city' alienated from the 'city of citizens'? Does global finance threaten global ethics? Or is reality more complex? If we knew more about the relations between financial technologies, distancing, and morality, we could contribute to understanding the paradox of distance and proximity raised by current 'electronic' teletechnologies *and* have more insight into the ethical, social, and political aspects of contemporary finance and its technologies. In particular, we would have more insight into the conditions under which more responsible financial practices are possible.

In light of the pressing societal problem sketched in the beginning of this chapter and considering the more general issue concerning moral distance raised by teletechnologies in the information age, there is an urgent need to inquire into the social and moral geography of current financial practices and their technologies. This book is written to meet this need.

## 1.2. Positioning in the academic landscape

Before I give an overview of the content and structure of the book, let me further position my approach in the academic landscape. I already said something about my approach, for example, about ethics of finance, philosophy, and interdisciplinarity. Let me expand this part for readers interested in the literature-geography of this book and its themes.

An obvious starting place for thinking about normative questions concerning financial technologies is business ethics, and in particular ethics of finance. Ethics of finance is a relatively young academic field. Although business ethics has always given some attention to finance, it is only during the last 20 years that it has grown into a separate subject in research and education. The field has rightly questioned the view, often taken by people working in the financial sector, that ethics is already being addressed by law and regulation: not all ethical issues are settled by law or self-regulation, and indeed not all ethical issues *can* be settled by law and regulation.[1] Furthermore, economists have largely ignored the topic. Instead, authors such as Boatright (1999; 2010), Kolb (2010a), Dobson (1997), and Heath (2010) have examined and questioned the ethical implications of assumptions in economics and finance (e.g., that rationality is about pursuing material advantage *ad infinitum*, that economic agents are purely self-interested utility maximizers, or that risk taking has always great value), have questioned and analysed the fairness of financial markets, and have discussed the ethical roles and responsibilities of finance professionals and the organizations and clients they work for (i.e., investment banks, pension funds, etc.). Some authors have also responded to the financial crisis with explanations and lessons (e.g., Kolb 2010b) and sometimes with an ethical analysis. For example, in his book on the ethics of investment banking, Reynolds (2011) argues that investment banks have specific ethical duties to their clients and that it is not enough for these banks to have Codes of Ethics, since these are mainly written to protect shareholders. Instead, he proposes (a new form of) self-regulation that deals with ethical issues. He also repeats the message that ethics is more than law and regulation, that 'compliance' is not enough: 'One lesson of the financial crisis has been that strictly legal behaviour, where it has ethical flaws, may nonetheless damage institutions, their employees and their shareholders. Actions may, while being strictly legal, also be plainly unethical' (Reynolds 2011, p. 6).

While ethics of finance and related forms of 'applied ethics' offer valuable insights into key ethical issues raised by contemporary finance, they do not give much attention to the role of *technology* in financial practices and its relation to ethics. Therefore, in order to tackle the specific questions concerning *technology* asked in the beginning of this introduction and in order to meet the objectives of this book – to understand and evaluate financial *technologies* – we need other theoretical resources and approaches. In order to complement existing work in ethics of finance with original work that brings out the specific role, nature, meaning, and ethical, social, and political significance of financial technologies, I will build on the state of the art in *philosophy of technology*, *social studies of finance*, and related studies of finance in *anthropology* and *geography*. These disciplines may be less known to many academic philosophers, but their approach(es) and work are exactly what we need in order to achieve this book's aim.

Let me start with philosophy of technology. After a century of systematic philosophical reflection on technology and its relation to ethics and society,

1 One could argue that there is even a *tension* between legality and morality. However, I will not further discuss this larger philosophical issue here and merely mention Reynolds' view.

philosophy of technology has developed into a subdiscipline of its own that can make helpful contributions to ethics and, more generally, to practical philosophy. In particular, it offers an interesting non-instrumental understanding of technology: propelled by the work of Heidegger, Ellul, Adorno, Horkheimer, Marcuse, Habermas, Winner, Feenberg, Latour, Bijker, Ihde, Borgmann, and others, the view has emerged that technology is never ethically or politically neutral, that it is not a neutral tool but that it has normative significance and societal implications, including *unintended* consequences. Phenomenological and hermeneutical approach to technology, for instance, have shown that technologies shape the way we see the world and influence how we live (together). Furthermore, there is a trend in philosophy of technology that moves away from reflection on technology as such to ethical reflection on specific technologies (Franssen et al. 2009). This has opened the door to empirically informed approaches that focus on the precise meanings and consequences of specific technologies and technological actions (e.g., engineering design) in many fields. The empirical turn has also facilitated collaboration with science and technology studies (STS), which studies the relations between technologies and society (for 'classical' works see, for example, Bijker et al. 1987 and Latour 1987).

This empirically oriented approach to philosophy of technology seems very suitable given the aim of this book to better understand and evaluate financial technologies: we need not only conceptual work – as practised by many twentieth century philosophers of technology *and* by many philosophers in the subdiscipline of ethics – but also *empirically informed* insights into financial practices and their social-material dimensions. Empirically oriented philosophy of technology and STS are therefore crucial elements in the approach of this book (see also below).

However, so far *philosophers* of technology have largely ignored the field of finance, including work in STS that has studied financial practices. This is regrettable, since much can be learned from so-called 'social studies of finance', which focus on the social and cultural aspects of finance and financial technologies. Again their approach is usually more empirical, taking distance from both theoretical sociological inquiry (for example, Marx, Weber, and Simmel) and financial economics. In the 1990s, STS pioneer Michel Callon started to apply actor-network theory to economics markets, and the past decade has seen studies of 'the concrete, material practices of trading' (Beunza et al. 2006, p. 721). Scholars in social studies of finance have studied 'heterogeneous assemblages of human beings and technical devices' in trading rooms (p. 722) or 'agencements' (Callon et al. 2007; see also Callon et al. 2005). They have also paid attention to the kind of knowledge traders and investors have and produce (MacKenzie 2005; Knorr Cetina & Bruegger 2002a; Beunza & Stark 2004), including their judgements (Beunza & Stark 2004), interpretations (Zaloom 2003) and the distribution of cognition (Beunza et al. 2006). Furthermore, they have revealed the '*locus operandi*' of contemporary finance (Beunza & Stark 2004, p. 372), that is, the spatial configurations of trading, and the embodiment (and again physical location) of the traders (Zaloom 2003; Zaloom 2006; Knorr Cetina & Bruegger 2002a; Beunza et al. 2006). (I will use and discuss this literature in Chapter 7.)

Given this book's focus on the relation between technology and distance, these studies about materiality, knowledge, and place in financial practices seem particularly relevant, especially if combined with (other) insights from economic sociology, anthropology, and human geography about *money*, which is arguably *the financial technology par excellence* and which will receive much attention in this book. For example, Davies and McGoey have argued that ignorance about the risks taken within the financial sector is actively exploited and used as an alibi by actors implicated in the financial crisis (Davies & McGoey 2012); Clark and others have argued that finance has become a media event and even 'entertainment' (Clark et al. 2004); Zelizer has done interesting research on the social meaning of money (Zelizer 1989 and subsequent work); and more generally there is plenty of anthropological research on the social roles and meanings of money (for an overview, see Maurer 2006), not only in so-called 'traditional' or 'primitive' societies but also in our contemporary societies. Financial markets have become 'the new exotic' (Maurer 2006, p. 18) and – as the social studies of finance literature also show – ethnographic methods are now being used in this area. Especially interesting with regard to the problem regarding distance and money is the question of how 'virtual' and how 'material' money is and if we should hold on to such a dualistic ontology at all. This question is also raised by literature in the field of human geography, especially cultural geography and of course *geography of money and finance*. For example, MacKenzie has argued that virtuality is materially produced and that global financial integration has not brought the end of geography (MacKenzie 2007). And in the literature on 'geographies of money and finance' (for an overview see Hall 2010) attention has been paid to technologies and the role of space and place (see again the literature in social studies of finance) as well as to political economy concerns about the geographical structure of markets and 'the spatial and temporal logic of global capital flows' (Clark 2005).

However, these approaches are not sufficient for a comprehensive and systematic understanding and evaluation of financial technologies (and hence for this book) for at least two reasons.

First, these (sub)disciplines and (sub)fields of research (STS, social studies of finance, anthropology of finance, geography of finance) lack a strong normative orientation. While the work done in these fields is never value free (which is sometimes acknowledged),[2] as social sciences they do not usually *intend* to bring out normative problems and lack the full palette of philosophical resources to articulate and discuss these problems in an elaborate and systematic way. But this is needed if we care to develop a better understanding of what I called the problem of *moral distance*, which concerns the *normative* dimension of the geography of finance.

2 In fact, scientific work is *never* value free, as much research in STS has shown.

Second, in so far as they have an empirical orientation, these disciplines do not always fully benefit from work in theoretical sociology: they often reject formal and conceptual analysis of technology and modernity, but I believe such work is still relevant and necessary to fully understand the relations between technologies, ethics, and society. In particular the work of Simmel (but also Marx and other 'classical' authors in sociology) is useful for reflections on the ethical, social, and political significance of money and other financial technologies. The same is true for more abstract, conceptual work in philosophy and for philosophical traditions such as critical theory, phenomenology, and pragmatism: such work is also very valuable for my purpose. The point is not that these theories are entirely absent from the discourse about financial practices and technologies in the mentioned (sub)fields and disciplines, but rather that the interpretative and normative force of these theories is not always used to elaborate the conceptual-reflective part of the research in a way that could deepen our understanding and support our evaluations of the practice. Note, however, that I avoid the claim that the disciplines in question should become 'less' empirically oriented; rather, I believe that in transdisciplinary research we should further *integrate* conceptual-theoretical work and ethnographical and other empirical studies.

This book takes such a transdisciplinary approach. We need both conceptual and empirical approaches if we want to tackle the hermeneutic and ethical problems raised by financial information and communication technologies. In order to better understand and evaluate financial practices and get a grip on the problem of the relation between financial ICTs and the various distances they may create, we need to bring together approaches from philosophy (including ethics of finance and philosophy of technology, but also other philosophical subdisciplines such as political philosophy, metaphysics, and epistemology – in each case both more theoretical and more empirically oriented approaches), from theoretical sociology, and from social and cultural studies of finance, including STS, social studies of finance, anthropology of finance, and geography of finance.

Such a multidisciplinary or transdisciplinary approach is always 'risky', in the sense that people looking at it from a strictly disciplinary perspective may not always sympathize with the way I use 'their' work or approach 'their' field, and also in the sense that specific transdisciplinary gestures may turn out to be less fruitful than others. But if what I do in this book is to contribute to coping with real-world problems it is well worth taking this risk. More generally, I believe only this kind of research can have a significant impact, not only within the academic community (in the fields of ethics of finance, philosophy of technology, social studies of science and technology, etc.) but also in a wider circle of people who think about finance and about the role of technology in society.

Finally, it is good to note that this book offers only a contribution to such a transdisciplinary project; it is at best and at most one small stepping stone towards understanding and dealing with the challenges we face in finance and in society. In the light of technological change, further inquiries into the ethical and social dimensions of finance are much needed – now and in the future. We all have a stake in it: financial professionals such as traders and bankers, of course, but also all people affected by these activities – and in a globalized world this means: everyone. It is reasonable to expect that, due to the rapid changes in the development of electronic technologies, the problem addressed by this book is likely to remain relevant, or even increase in relevance. The same is true, more generally, for (other) teletechnologies and the worry that they create various kinds of distances. In the foreseeable future, there is likely to be a further widening of the gap between, on the one hand, the pace of technological development and, on the other hand, our capacity to understand, evaluate, and steer its course. It seems also likely that globalization is here to stay. What kind of technological world are we building, and what kind of technological world do we want? More research on the ethical and social aspects of technologies – financial technologies and others – will be much needed in order to make sure that we keep a grip on what is going on, and that technologies designed to bring us together do not drive us apart.

## 1.3. Content and structure of the book

This book constructs and critically discusses a story of how financial technologies have contributed to various kinds of distancing, including what I call 'moral distance'. Digging into the history, metaphysics, and geography of money, algorithms, electronic currencies, and other financial technologies and artefacts, I first argue that while financial ICTs bridge physical distance between people, things, and places, they also risk rendering relations to others, to ourselves, and to the world less personal, more disengaged, placeless and objectifying. As such they *create* distances of a social, moral, and epistemic kind.

I start this argument in Chapter 2, in which I briefly construct the history of financial technologies and interpret it as contributing to, and being part of, the rise of various kinds of distances. I show that there is a relationship between the financial technologies that emerged in ancient agricultural civilizations and the increasing social, political, moral, and epistemic distance in these societies, and that financial technologies also played a key role in globalization processes already in the ancient times. I say more about the history of money and show that the history of its dematerialization and its distancing and disengagement effects already started in the ancient world. But I also pay attention to financial technologies which usually receive less attention but are an important part of the history of financial technologies and distancing: writing, the invention of numbers, military and mining technologies, technologies for recording and accounting such as clay tablets, printing techniques, taxation and management methods, and of course calculating devices and other 'money machines' – there is a history running from the abacus to contemporary computational finance. I argue that these technologies create distances: physical-geographical distances, but also a distance to others, a remote kind of morality and responsibility, and even distance to oneself. Using Foucault I make a point about the mode of surveillance and control/power in contemporary society.

In Chapter 3, I further explore the relation between financial technologies and various kinds of distancing by drawing on Simmel's work. I analyse and discuss his arguments about money and connect these to a current in philosophy of technology that also draws our attention to the alienating and disengaging influence (or potential) of technology. In particular, I focus on Simmel's view of the mediating role of money in trade, his characterization of money as a 'pure tool' that dematerializes and creates abstraction, and his arguments concerning the social role of money: how it objectifies, de-personalizes, detaches, creates distance, and alienates. I show how these views are in line with those of some contemporary philosophers of technology who emphasize the relation between technology and disengagement (Dreyfus, Borgmann). I also critically discuss the application of Simmel's theory to contemporary financial technologies by using more recent literature in anthropology of money and finance, and discuss some problems with Simmel's view. I make brief comparisons with Marx's view and McLuhan's view where appropriate (I say more about Simmel and McLuhan in Chapter 4).

In Chapters 4 and 5, I discuss the question concerning the nature and meaning of money and other financial technologies in a global context and in the 'electronic' or 'information' age. Are streams of information only 'global'? What does money do with space and time? What is the nature of electronic money, for example, Bitcoin?

Chapter 4 introduces questions regarding the geography of money and financial technologies. Using McLuhan, Arendt, Castells, and others, I first study the relation between globalization and technology, in particular electronic ICTs, and suggest that what Castells calls the 'global space of flows' has various distancing effects and implications for social relations and responsibility. Then I turn to the relation between *financial* technologies and globalization. How do we experience global finance, and what are the consequences for society and culture? What is 'message' of money as a medium in the information age? How does it shape our experience of space, our social life, our 'inner' life? Reading McLuhan and referring back to Simmel (and Marx), but also using more empirically oriented literature on contemporary finance, I articulate some darker and brighter sides of financial globalization. I identify again distancing effects and the danger of distancing and alienation, but I also point to the social and moral promises of the new financial technologies.

In Chapter 5, I discuss the meaning and 'metaphysics' of money, with a particular focus on electronic monies and currencies. I reflect on the concepts and distinctions we use when we think about electronic money and I discuss the nature of Bitcoin. First I distinguish between several 'ontologies' of money (object ontology, information ontology, social ontology), which each have implications for the nature and meaning of electronic money and electronic currency. I show how the object-ontological approach is challenged by money and especially the new electronic forms of money, discuss the merits and problems of an information-ontological approach, respond to Searle's social ontology of money, and offer what I call a 'deep-relational' view of money which changes the ontological question into an epistemological, social, and phenomenological one. I argue that what money 'is' and what other financial media and technologies 'are' is not observer-independent and instead is shaped by human subjectivity, language, and social relations. Then I further discuss the relations between money and technology, and between money and (social and political) institutions. I argue that money is a technology and medium, but then with 'technology' properly understood as being more than a means to an end; instead, money and other financial technologies transform our human and social world. They are not mere means but also shape our ends. Financial technologies and other technologies even transform what it means to be human. This means that they can create various kinds of distancing, but also that they can contribute to a transformation of our relations and our social and political institutions (see also Chapter 8), including supporting new forms of trust. I give the example of Bitcoin and its independence from (trust in) the nation state.

In Chapter 6, I turn to the epistemic and moral implications of the new electronic technologies in global finance. I introduce these new technologies and practices and emphasize that neither ethics of finance nor philosophy of technology (ICTs, AI, robotics and automation) have paid sufficient attention to them. Then I discuss how the use of algorithms and other 'money machines' in contemporary global finance re-shape (conditions of) *responsibility*. What does it mean for moral responsibility when trade becomes increasingly dependent on ICTs? Is it still possible to exercise responsibility when trade becomes *automated*, when 'the machines take over'? Is financial AI moving us towards a 'financial singularity'? I also consider the use of screens and other electronic ICTs, but focus on the case of high-frequency trading (HFT). I argue that in the context of electronic globalization, financial ICTs threaten the fulfilment of two basic and classic conditions of moral responsibility: both experts and lay people increasingly lose *control* over what is happening in global finance, and lack sufficient *knowledge* of what they are doing. For instance, HFT can be interpreted as a kind of financial 'auto-pilot' which raises the question of who is responsible when the algorithm is malfunctioning and creates chaos on the market(s). And the new technologies increase the distance to other traders and to people affected by their trade and investment decisions, and further limit the knowledge the general public has about what is happening in HFT and global finance more generally. Again there seems to be a development towards various types of distancing in global finance, to which the new financial technologies and media contribute. How can we cope with these problems as professionals and as citizens?

Yet the book also shows that the heavy use of electronic ICTs in global finance in no way means that it has become a 'virtual' or non-real world. On the contrary, in Chapter 7 I question and nuance the arguments in the previous chapters by 'localizing' and 'materializing' finance and financial technologies. I draw on social studies of finance and technology to show that its technological practices are very material and place-bound: the geography of contemporary financial experience and action is less 'global' and 'virtual' than one would expect; the local and the material remain important. For instance, people's financial situation depends on where they live (there is a 'digital gap' and a 'financial gap'); 'global' finance continues to have 'local' forms, dimensions, and impacts (see, for instance, the concept of 'glocalization') and takes place at specific financial hubs; the social space of traders also has a local orientation; and ICT-mediated trading depends on material infrastructures. STS and social studies of finance help us to reveal the 'material world' of financial practices: the networks of humans and non-humans that produce financial knowledge and action, the screens and the bodies, the trading floors and the cockpits of techno-finance. Technologies such as HFT algorithms and the Bitcoin algorithm turn out to have important material dimensions. Moreover, the consequences so-called 'virtual' technological-financial practices have for the conditions under which we ascribe and exercise responsibility, the influence they have on our lives, and the social impacts and social vulnerabilities they create are all very real indeed. Furthermore, I argue that global finance also has 'human' and 'personal' dimensions: there is room for personal relations and for human judgment and interpretation. Perhaps there are also more personal forms of money and there is a sense in which cultures and practices can get re-humanized and re-personalized. If financial-technological practices are human through and through and if distancing is shaped and responded to in concrete social-material practices, then this opens up a space for social-financial change. I argue that we are not helpless victims of a determinist and fully autonomous technological development, and that the best response to the distancing problems we face is not to reject 'technology' but, among other things, to change the financial technologies and media we use.

In Chapter 8 I further discuss this issue of financial-social change and explore alternative financial technologies, institutions, and practices. I ask how the darker side of contemporary techno-financial developments can be resisted, or better, how they can be re-oriented into a socially and morally responsible direction – not against technology but *with* technology. In the course of my discussion I also pay some more attention to the issue of power. First I show that we can try to *resist* developments towards more epistemic, social, and moral distance. For instance, we can identify uses of financial ICTs that increase social distance between 'haves' and 'have nots', and between 'global' financial players and 'local' people who have far less power. For example, we may want to resist financial-technological systems that produce, for instance, food alienation. But we can also try to frame the problem (and hence the solution) differently. If financial technologies are part of the problem, then they are also part of the solution. By thinking about, and experimenting with, new financial technologies, we can contribute to the growth of *alternative* financial practices which may help us to 'un-distance'. This approach is more future-oriented and constructive, and attends to positive changes that are already taking place. New financial and trade technologies promise to bridge or reduce epistemic, social, and moral distancing. If they succeed, they also may help to create conditions under which it is easier to exercise and ascribe responsibility and thereby effectively support more responsible finance, trade, production, and consumption. After reviewing some new social-financial-technological experiments and showing how they may reduce epistemic and moral distance (alternative trade practices such as fair trade and farmers markets, alternative local and global currencies such as LETS, Bitcoin, money in games and virtual worlds, and alternative financial systems and movements such as micro-credits and Slow Money), the book ends on a cautiously optimistic note and asks us to try out and imagine new financial technologies and practices that re-shape our current relations to others, to ourselves, and to the world.

In my conclusion (Chapter 9), I first summarize and conclude my discussion about financial technologies and distance, with a focus on the issue of social change, and then I explore some implications for (1) academic work in philosophy of technology, (2) responsible and democratic research and innovation in the field of ICTs, and (3) our lives.

# Chapter 2
# From clay tablets to computational finance: A brief history of financial technologies and distancing

## 2.1. Introduction

In this chapter, I offer my interpretation of the history of financial technologies, in particular the history of money, in order to start developing my main thesis about the relation between financial technologies (in particular financial ICTs) and distancing. With the term 'distancing' I refer to the various kinds of moral and social distances made possible by technologies that were meant to bridge physical distance, that were meant to mediate between people and between people and things. My method is not to provide an 'objective' or 'neutral' history of 'facts', but quite the opposite: to use literature on the history of finance in order to build empirical and hermeneutical support for a conceptual argument about financial ICTs and distancing, which I will further develop in the next chapters and which I will critically discuss. The main aim, therefore, is less about 'getting things right' and more about contributing to 'making sense' of money and financial technologies and about exploring and evaluating their influence on our lives and our culture. It is about getting insights into the phenomenological, social, and moral aspects of financial technologies. My question is: what is the influence of money and other financial technologies on our relations to reality, to others, and to ourselves? I will construct the history of financial technologies as part of a process of increasing epistemic, social and moral distancing, thus contributing to more insight into what I call the 'moral geography' of money and other financial technologies. I will focus on specific transformation processes and critical changes in the material-cultural history of our culture and indeed our civilization and on the role of financial technologies and artefacts such as money, numbers, and writing in these processes and transformations in order to discuss the epistemic, social, and moral significance of these processes and these technologies. I will discuss the history of money, of course, but I will also pay some attention to the history of other financial technologies. This exercise will prepare the ground for my next chapter, in which I will further conceptualize distancing and 'moral distance' and further build up support for my argument about it using sociology (Simmel and others) and philosophy of technology (mainly Heideggerian philosophers: Borgmann, Dreyfus).

Let me start in this chapter with the financial technologies that emerged in the ancient agricultural civilizations. Then I will say more about the history of money and other financial technologies.

### 2.2. The birth of financial technologies, ancient bureaucracies, and the issue of distance

The history of financial technologies is a fascinating one which does not always get the attention it deserves, but which needs to be studied by everyone interested in the philosophical and social study of contemporary culture and technology. Perhaps the lack of attention from humanities and social sciences stems from the view that it is a mere 'technical' subject, only of interest to finance scholars or to a few historians and archaeologists. But this view is mistaken. Money, for instance, is not only about coins or payments; it is about trust and relations, about social structure and identity, about innovation and work, about politics and how we organize societies, about culture and civilization, and about how all this is changing and might change in the future. It is about how we do things, about how we think, about who and what we are and may become. Thinking about the history of financial technologies, therefore, is not just thinking about artefacts such as coins or banknotes, or about the institution of 'money' understood as something separate from all our other activities and interests (if that makes sense at all). Rather, it is thinking about the history and future of our culture and civilization, about the human, about ourselves.

Consider the story of what we may describe as 'the birth of money' but also 'the birth of writing', 'the birth of counting and accounting', 'the birth of computing', and even 'the birth of bureaucracy'. The general story – some authors may call it a 'myth' (see below) – runs as follows. Before money was used, there were gifts, there was sacrifice, there was local distribution of goods. If there was trade, it was barter. Goods were exchanged without the mediation of money. In a gift economy, there was not even an agreement on a return. Perhaps there was a sense of 'I owe you something', but this was not formalized. There may have been rituals for barter, which could have an impersonal and formal aspect, but the rituals were part of a culture and an economy in which there was a high degree of trust. The barter system presupposed trust in each other and a tight social structure. People knew one another, had close relations with one another and generally they also had direct, face-to-face, contacts with strangers. People lived in relatively small groups and communities. Moreover, as Roberts says, these exchanges were not done for making profit and exchange ceremonies were marked by attitudes of respect (Roberts 2011).

Things began to change dramatically, however, with the introduction of agriculture, that is, with agricultural technologies and the related social changes they helped to bring about. In the beginning of agriculture there was not yet money, but financial technologies started to develop, together with social innovations. My understanding of this process is that agriculture enables people to 'have', keep, and possess goods, and that this has a lot of social and technological conditions and implications. In a hunter-gatherer society, and also in early agricultural societies, goods are usually for immediate consumption. There is little or no 'stock'. And if, as Roberts argues, there is not much to distribute, there is a lot of sharing, there is nearly equal distribution – even if there were traditional differences in status, ceremonial priority, and so on (Roberts 2011). But once agricultural techniques get better, there is more than people need for immediate consumption, which also raises questions about who 'has' what, about who controls the goods. It is possible for some people to exercise power over others, not as the chief of a tribe, but as the lord and master. Political changes followed techno-agricultural ones. Roberts argues that, in particular, the development of irrigation systems made it possible to cultivate larger areas of land. Priests and kings started to claim the best land and forced farmers to work on it. Temples were turned into the ancient version of factories (Roberts 2011). And when there was enough food, there were also demographical changes: cities emerged.

Moreover, if you have a stock, you need to count and 'keep stock'. You need to calculate. You need to write down what you have. If you tax people, you also need to calculate and keep track of numbers and quantities. The agricultural revolution thus made possible – and one could say: made necessary – new social forms of organization and new technologies, including financial technologies. The agricultural revolution is also the birth of financial technologies: technologies that help you to keep stock, to write down what you have or what you owe to others, to manipulate and control other people, to calculate, to administrate. This is not only an economic or financial change; it is a change in the way we experience the world and treat others. It is also a change in the way we understand ourselves: we come to see ourselves as managers of people and goods, as bookkeepers and stock-keepers, as accountants of other entities.

In *Money: The unauthorized biography*, Felix Martin rightly calls ancient Mesopotamia, usually seen as the cradle of agriculture and (hence) civilization, the 'ur-bureaucracy' (Martin 2013, p. 37). It was not only a birthplace of the city but also of new geographies and forms of social organization such as the city and hierarchical social structures with kings and priests ruling the people. Similar forms of social organization *and* financial technologies developed in Egypt, China, and other parts of the ancient world. For instance, up to 4400 BC the people in Egypt lived in 'egalitarian, tribal arrangements'; by 3000 BC 'this tribal society had been transformed into an agricultural society' (Farag 2009). This meant that forms of supra-tribal control emerged (and later non-tribal control) and that larger territories had to be managed. Large numbers of people needed to be managed as well (e.g., to build the pyramids). The bureaucracy grew. In Egypt, the pharaoh (the term is related to the word for palace or domain around the temple) had accountants to write down which goods were taken. Tribal obligations had been replaced by taxes. The vizier exercised authority in the name of the king or pharaoh. He supervised the scribes, the tax assessors and tax collectors. Priests also played an important role in the process, in particular in the (re-)distribution. All this required 'the development of an elaborate accounting system' (Farag 2009).

Such a system required different, new technologies. It seems that writing was at least partly, if not mainly, invented for the purposes of accountancy and taxation. It was, among other things, a *financial technology*. Early writings were found in Uruk (Mesopotamia), Egypt, China, and Columbia (Aztecs). An important, later text in the history of early writing is the Code of Hammurabi, a law code used by the ancient Babylonian king of that name which dealt with, for example, debt. Around the time that code was written (*c.* 1760 BC) money had been already invented, but the links between finance/accountancy/administration, technology, and culture were again very strong. Everywhere in the ancient world there was a need to write down quantities for the purposes of bookkeeping, administration, and trade. These purposes required the invention of writing in general and, related to that, the invention of writing systems (e.g., the Phoenician writing system and later the Greek alphabet) along with the invention of numbers. These inventions were once again important milestones in human culture, including in the history of mathematics and its links to abstract thinking, and in the history of finance.[1] Furthermore, bookkeeping or accounting can also be considered a technology. Martin writes:

> The hierarchical control of economic activity by clerical bureaucracies required a management information system: a technique for quantifying stocks and flows of raw materials and finished goods, for using these quantities in forward planning, and for checking that the plan was being correctly carried out on the ground. Accounting was a social technology that combined the ability to keep records efficiently using writing and number with standardised measures of time so that quantities could be tracked as stocks on balance sheets and flows on income statements. (Martin 2013, p. 43)

In the ancient world, the development of new financial technologies thus went hand in hand with the agricultural revolution and its associated revolutions in political, religious, and social organization. The same close ties with larger culture and society can be observed with the birth of money. But before I say more about money, I would like to offer a further interpretation of what happened in the ancient world when ancient technologies were invented together with new social-cultural developments. It strikes me that the agricultural revolution was also a geographical, social, and moral revolution that can be interpreted as a 'distancing' revolution and as contributing to 'the birth of moral distance'. The financial technologies mentioned here were not only artefacts or techniques, they were also strongly related to new social and geographical relations. This change was of moral significance. As the small groups and communities were replaced by cities

---

1 See, for example, relation between Fibonacci's mathematics and financial transactions in his day (e.g., as discussed in Ferguson 2008, p. 31–32).

and kingdoms, there was not only more physical but also more social and *moral* distance between people. The 'management information system' started to replace relations of trust with 'accountability'. The management technologies were also at the same time distancing technologies. They bridged the physical distance between the palace and the people 'on the ground', but they also made possible larger territories (thus increasing the physical distance in another sense), bridged but also *created* a knowledge gap (it became more difficult for the ruler and the people to know what goes on in the palace or 'on the ground'), and increased the moral and political distance between people. The gap between the palace (the king) and the people (the king's subjects) increased. The technologies made possible a society in which there was more inequality, more 'cold' or 'business-like' relations, less 'community'. People lost full control over their activities; they became part of a larger society, a larger social machine. They became managers or were directed by managers.

Thus, if today we complain about modern bureaucracy and its influence on relations between people (e.g., employees), it is not only the 'modernity' that is to blame. The development towards more 'remote' technologies, social relations, and political structures already started in the ancient world. Moreover, the way this 'moral distance' came about was largely unintended; it is and was the consequence of structural material-cultural and material-social changes, in which technologies play a key role. Our contemporary world of management and accountancy has its roots in that ancient invention of new technologies and new social relations. Today we might no longer have powerful kings and priests. But the managers and the accountants are everywhere, and the type of relations that emerged in the world of the ancient kings and pharaohs are unfortunately still in place. We now see the world as what Heidegger called 'a standing-reserve', which we can manage and use for our purposes. This way of looking at the world and indeed at others (and at ourselves – see later in this chapter) has reached a climax in modernity, perhaps, but a milder version of it already emerged in the ancient world. It is good to keep this in mind when reading the next chapters, in which I will say more about modernity. But let us first take a closer look at the emergence of the financial technology *par excellence*, money, and how it contributed to the history of distancing.

## 2.3. The birth of money, its dematerialization, and the emergence of banks

The history of money starts with money as a medium of exchange used in trade. A simple form of barter works as long as both parties want each other's goods. But this 'double coincidence of wants' does not happen very often. Moreover, for barter to work well there needs to be equality of value. But often equality is difficult to achieve or it may be even unclear. And if it is not clear, how can it be measured? In his history of money, Morgan has argued that at an early stage of human economic activity, a standard means of exchange was found against which the value of goods could be measured. In many cultures, this standard was cattle, but it could also be cloth or cereals (Morgan 1965, p. 11). The Babylonians used silver, lead, copper, bronze, and gold, but also honey, oil, sesame, wine, papyrus rolls and arms. Ornaments were also used as a means of exchange, and some valuable objects circulated in large parts of Asia and Africa; apparently there were already widely shared standard means of exchange long before modern 'globalization' happened. For example, cowry shells were used in Africa, India, and China. Later metal coins became predominant, often with the seal of the ruler who guaranteed its value.

The precise relation between money and social organization (and moral relations), however, is more controversial. On the one hand, money promises freedom from centralized organization and oppression. Martin argues that money emerged in ancient Greece and in Asia Minor. Propelled by what we would now call the 'modern scientific world view' (Martin 2013, p. 56) and universalist, more abstract thinking (here: the idea of universal economic value, but also the emergence of the gap between a distant observer and an objective universe), the Greek world managed to move away from traditional social structures and obligations. Money was a means to become free – at least if you possessed it. Morgan even suggests it might have played a role in the decline of slavery (Morgan 1965, p. 14); money thus became a freedom tool. More generally, money makes possible different relations between people. Trust is no longer needed at all, at least not trust in the sense of *personal* trust; it no longer matters from whom you get it. A more impersonal – but also freer – relation is possible. On the other hand, these different relations also become more 'remote'; if trust is no longer needed it also means that the relation becomes more impersonal. And the introduction of money may make people free in one sense, but not necessarily in another. For example, money as such does not necessarily liberate you from hierarchical power. In fact, often quite the contrary happened. Rulers used money as an instrument of rule (and often oppression). As Morgan points out, governments still played an important role in the money system: it was the government's seal which acted as a guarantee. More generally, money was also used as a tool to exercise and consolidate power. For example, Alexander the Great installed a unitary monetary system in the regions he conquered. Furthermore, money also played and plays a role in the organization and control of the work of others (i.e., by paying workers). And as Marx argued, when work is turned into a commodity, the worker is alienated from the work and from others. It also gives some people with money (capitalists) the power to control and exploit others (those without money).

In both interpretations, however, money plays a key role in increasing social, epistemic, and moral distance. In the course of the history of civilization, relations of trust were further eroded, scientific-style thinking created distance from the world, and obligations were no longer linked to 'local' or particular personal relations but became impersonal, universal, and potentially global. Money was an amazing tool: it could bridge significant physical distances and made possible what we might call proto-global trade. But it also played its part in rendering epistemic and social relations less direct, in promoting impersonal rather than personal relations, and in making possible a less direct relation to, and engagement with, reality.

Let me say more about this epistemic aspect and its link to social and moral relations. The change concerns the directness and remoteness of experience and knowledge. If you engage in barter, you know the goods you are trading, you know their value, and you know the people you are trading with. In particular, you know the value of, for example, *your* cow and you know its value and its worth *for you*, that is, you know its meaning and value in relation to your personal life and your personal activities. When the cow becomes a means of exchange rather than a good in itself or indeed 'an animal', there is already more relational distance. There is no longer a direct relation to the goods and to the people, since with a means of exchange more remote trade is now possible. But with the universalization of money, it is possible to trade without *any* direct relation and personal engagement with things, animals, and people. Both the epistemic and moral relation to reality and to others has become more distant. Whether or not this is 'freedom' and 'liberation', and, if it is freedom, whether the 'cost' of this freedom outweighs the 'gains', is questionable. (I will say more about this question in the next chapter).

Furthermore, we can also see in the history of money another kind of *distancing*: the widening of the gap between the material value of money and its monetary value. This is the history of the progressive dematerialization of money, that is, the history of an increasing distance between money and matter. The first means of exchange such as cows and shells were still very physical, and the first forms of money – coins – were very material indeed. They were substantial, metal artefacts. However, in the course of its history, money became less substantial and less material. In the Middle Ages money was already clipped and counterfeited. For example, the weight of the pound was reduced (Morgan 1965). And whereas in Anglo-Saxon England it was the equivalent to one pound weight of silver, later sterling silver was used (hence 'pound sterling') and in the sixteenth century a coinage was used that contained only one third silver and two thirds copper. The material value of the coins thus decreased over time. Around 1700 the Bank of England and the Bank of Scotland introduced paper money. The monetary value of money became, in a sense, what it always had been but was now more fully revealed: a matter of social acceptance and trust, often (but not necessarily, see for instance Chapter 5 on Bitcoin) guaranteed and sanctioned by a central authority. Ferguson writes:

> Money is not metal. It is trust inscribed. And it does not seem to matter much where it is inscribed: on silver, on clay, on paper, on a liquid crystal display. Anything can serve as money, from the cowrie shells of the Maldives to the huge stone discs used on the Pacific islands of Yap. And now, it seems, in this electronic age nothing can serve as money too. (Ferguson 2008, p. 30)

I will return to this last issue in Chapter 5 on the metaphysics of money. But let me focus here on Ferguson's message that money is about trust. However, with the development of money, it was no longer personal trust that counted; it became trust in the system.To understand this change, and to fully understand the introduction of paper money and its social and moral consequences, however, we should not only look at money in the sense of artefacts, but also at the social institution of debt.

Debt plays an important role in the history of money. According to many contemporary anthropologists, archaeologists, and economists, debt and the gift (which also creates a kind of debt) even emerged *before* money (Mauss 1925; Hudson 2004; Hart 2005; Graeber 2011). They call the story about the barter origins of money a 'myth'. However, while this alters the story told in the beginning of this chapter, it does not really change my interpretation of the history of financial technologies as a history of distancing: if one rejects the barter 'myth' and starts instead from the debt origins of money story, then one could say that first there are more (inter)personal forms of debt, whereas later there are more distant social-financial relations: relations to the temple (priests) and the palace (the king), and later relations to banks and nation states.

Debt is about trust, too. As Ferguson says: it is no coincidence that the root of 'credit' is the Latin word *credo*, the Latin for 'I believe' (p. 30). The history of debt is therefore also a social and political history. But in this history, financial *technologies* and the specific forms they take play an important role and has implications for trust and distance. What debt was changed significantly with the introduction of new money and other new financial technologies. Before new financial technologies developed, there was 'debt' in the sense that people felt the personal obligation to give something back to another person. With new technologies such as writing and paper, however, it became possible to *formalize* debt. Already in ancient Babylon (3000 BC) temples took deposits and gave loans, and thus acted as banks. Much later we see emergence of paper money and the birth of banking as we know it. Although paper money was already used in eleventh century China, it was in (late) medieval Italy and Flanders that money traders started using promissory notes in order to facilitate transport over longer distances. For instance, Galbraith mentions Renaissance Italy as the origin of banking (Galbraith 1975). Moreover, as Morgan tells, goldsmiths also played an important role in the development of banks and bank notes. In sixteenth- and seventeenth-century London, goldsmiths began to act as a bank: they issued receipts in the form of a promise to repay to a named depositor. Credit emerged in this way: first the receipts were a guarantee of the gold kept in the smith's place, then the smith realized that he could issue more receipts than the amount of gold he kept. Later the phrase 'or bearer' was added and receipts started to be circulated among people (Morgan 1965, p. 23); they were used as money. Later the Bank of England issued notes; governments started to control and later monopolize the issuing of paper money. However, until the eighteenth century one could still demand that bank notes should be paid out in gold, and for a long time there was a 'gold standard' as the basis of the monetary system. Although many (national) banks still keep gold reserves, this system is no longer in use. What remains is acceptance and trust, trust in an impersonal monetary system. This also means that, in principle, new currencies can be used and this is also happening (see Chapter 5 on Bitcoin and Chapter 8 on alternative financial technologies). Surely, it is always difficult for a new technology and currency to establish itself. For instance, today Bitcoin is still mistrusted by many people (including investors and speculators) and therefore remains 'unstable' as a monetary technology. But it is clear that the functioning of money as a technology and institution does not depend on the precise form it has. It can be coins, paper, or electronic money. What changes with the form, however, are the social and moral relations.

Indeed, the precise form of money matters, epistemically, socially, and morally speaking. Exactly *how* it is related to the social still depends on its form. Coins, bank notes, and electronic monies are all part of specific types of societies and make possible different kind of relations. Different financial technologies do different things to how we experience the world and how we treat one another. As suggested above, the progressive dematerialization of money seems to go together with the increase of social and moral distance. Personal trust gradually disappears and is replaced by different kinds of trust – if trust at all. Moreover, there is a clear relation between the history of debt and the history of financial technologies. This started already in ancient times. Morgan writes that the early records from Babylon, Egypt, and Greece consist of 'records of deposits and loans kept by temples or by secular money lenders; records of obligations to the state; and records of expenditure kept by public officials' (Morgan 1965, p. 36). This relation between debt and technology has social and political implications: debt establishes asymmetrical power relations: between citizens, between citizens and banks, or between citizens and political (or religious) authorities. Financial technologies help and always have helped to establish and maintain these relations, but also to challenge them (for example, with decentralized forms of money). And, as suggested before, they have also played a key role in the rise of accountancy, management, forecasting, and so on – and have therefore helped to make possible the accompanying social relations and (dis)trust.

Keeping in mind the current ethical challenges in finance, but also the historical power of bankers and the key role they played in the history of money, debt, and globalization (consider, for instance, the success of the Medici family in the Renaissance – they were 'foreign exchange dealers' and *banchieri* (Ferguson 2008, p. 42)), we also need to say something about banks, their sources of power, and the role technology plays in creating and maintaining that power. One reason why bankers were and are so powerful has again to do with 'monetary fiction' and trust, and with financial technologies. When goldsmith bankers started issuing banknotes in seventeenth-century England, they soon realized that they could provide more loans than the gold they actually possessed. This system works as long as not everyone demands their gold (or money). It is an issue of trust – trust not in specific persons but in the bank and in the entire monetary system and institution. But this only works because of the specific financial technology: paper money makes possible that there is a distance between monetary value and material value, between paper and gold. New technologies thus make possible new forms of banking and their power.

For the purpose of ethical and social analysis, there is also another distance that is very important, one that is more physical and geographical, but which is also of social and moral significance. I already mentioned that money makes possible larger territories and distances. And Morgan's point was that traders needed a technology (paper money) in order to make transport over longer distances easier. Let me now say more about the relation between the history of globalization and the history of financial technologies. (This also prepares Chapter 4 on globalization and contemporary financial technologies).

## 2.4. The geography of distancing: A history of globalization and financial technologies

Financial technologies make possible globalization and play a key role in its further development. Already in the ancient world, means of exchange circulated in larger kingdoms and empires, and beyond. In Mesopotamia more 'global' business was not only propelled by the invention of writing, calculation, accounting, forecasting and legal systems; it was also made possible by the larger size of the territories, for example, Hammurabi's kingdom (Roberts 2011). Since the territory was vast, Assyrians and Babylonians had to transfer money and credit over long distances. This process of globalization continued in later times. By the thirteenth century, there were international payment systems (Morgan 1965, p. 159). First merchants used to meet up in person. Later 'bills of exchange' were used (comparable with a cheque) and this enabled transactions over distance.

Already in the ancient world, trade was based on exchange of different resources. Like today some territories and their people had more precious resources than others. For instance, Egypt had water, fertile land, gold, copper, and building stone, whereas where good land was scarce people lived nomadic lives (Roberts 2011). Some societies also provided workforce. All these exchanges were bridging long distances, because of expanding territories. When business and finance flourished in ancient Greece, this was not only due to the invention of coinage but also to the large territories conquered by the Greeks: this meant again that trade had to cover long distances. Furthermore, the Greeks could rely on slave labour. The role of technology is important again: slave labour was related to military technologies and other technologies (e.g., military technologies meant that land could be conquered and people could be enslaved; mining technologies required workers, for which slaves were used.). Roberts argues that these conditions contributed to the development of an advanced culture, to sophisticated science, art, and philosophy, and to the political and legal structures of the city state. Indeed, it is good to remind ourselves that Greek philosophy and democracy, which we now value so highly, developed and flourished on *the market* square. Markets were spaces for business and financial exchange as well as for 'high' culture and politics.

In ancient Rome we find similar developments. Roberts claims that there too 'business originated in war' and 'flourished in peace' (Roberts 2011, p. 133), in particular in the peace of the Roman empire, whose size enabled the rise of gigantic economic and financial space which we may call 'proto-global'. Rome also had a very developed coinage system. Coins circulated through the entire Empire. There were *negotiatores* and *argentarrii* who acted as bankers because they lent money and kept deposits of money, such as in Athens during the second half of the fifth century BC (Andreau 1999, p. 30). Money lenders would set up their stalls on a long bench called *bancu* – the origin of the word 'bank'. Trade developed between military posts in the empire, and the rise of maritime trade in the second century BC made possible the mass exchange of commodities. Ships made possible trading higher volumes. Some provinces were specialized in particular goods, for example, grain in Egypt and wine and olive oil in Hispania and Greece. Tax was collected in the provinces; money was transferred. But because the sea was dangerous – storms and pirates – the Romans already found a means to execute money transfer over long distances without having to transport coins. It was not quite 'bank notes' they used (as explained above these had a different origin; the Romans did not have paper money) but a form of book transaction: what they did was accept payment in one city and arrange for credit in another. Again the technique of writing solved a financial issue, which is at root again also an issue of distance. Writing is a memory tool, as Plato already knew (see the argument against writing in the *Phaedrus* – 275a–b) and for example, philosophy and literature have benefited from this, but it might well have been a *financial* memory tool first. And like other forms of writing, there is the danger that writing in finance renders knowledge less personal, less vivid. There is the danger that personal experience, which according to Socrates was a necessary condition for wisdom, is replaced by things that are written down. For finance, this means that there is a danger that financial knowledge becomes alienated from personal relations and personal experience, that financial expertise becomes divorced from financial *wisdom*. Did this really happen? Is it really a problem? In the next chapter I will further inquire into the epistemic and moral impact of money and other financial technologies.

To end this section, let me emphasize again the strong connections we find between financial technologies and trading technologies, on the one hand, and the geographical and economic development of the great ancient civilizations, on the other hand. This link has been clear throughout this chapter, and further inquiry into the history of finance and economics only confirms it. Trade and finance in ancient Greece and in the ancient Roman Empire can only be fully understood in the context of larger transformations in the ancient Mediterranean economic space which made possible their flourishing. For example, between 1000 BC and 500 BC there was a social and economic transition which prepared the new empire-type worlds of Greeks, Romans and Persians. Susan and Andrew Sherratt list the following changes, which are relevant to the history of financial technologies and which show the link between technologies, economy, and geography: the transition from centrally controlled, bureaucratic trading to merchant enterprise; the production of iron; new forms of empires involving new forms of tax and tribute; the rise of a tension between commercial interests in independent city states versus the interests of territorial empires; the establishment of formal colonies which were territorially defined; different military organization and military technologies; consciousness of ethnic differences; slavery used for agricultural work and mining rather than only household slavery; the growth and intensification of mining since precious metals were needed for tribute; commerce and military expenditure; and the expansion of the territory and the economy (Sherratt & Sherratt 1993, pp. 361–63). This was the beginning of what the authors call 'a growing world-system' (p. 375), and it is clear from this list that financial-technological developments are intrinsically connected with other developments as they are embedded in, and result in, different social-geographical and political-geographical realities. In my chapter on globalization and financial technologies (Chapter 5) I will say more about the contemporary 'world-system' (that of the 'Information Age') and also reveal what the Sheratts call 'complementary aspects of an unfolding pattern with its own internal logic' (p. 361). We will see that today there are also new social and moral geographies that arise as part of an unfolding pattern in which financial technologies play not a marginal but a *central* role.

## 2.5. Recording devices, print technologies and calculators: The rise of money machines and its implications for our social relations, responsibility, and subjectivity

In the previous sections I mainly focused on money and related financial technologies. But there are also technologies that are not usually considered as 'financial' but nevertheless deserve to be discussed as such in this book since they are linked to finance and made forms of distancing possible. I already mentioned the role of writing (technologies) and the invention of numbers and other what we may call 'mathematical technologies'). I also pointed to links with military technologies and mining technologies. There are of course also links with the use and production of specific commodities (e.g., silk). Let me now say more about (1) the history of 'money machines' in the stricter sense of technologies used for the material production of money and (2) the history of technologies used for trade and bookkeeping. This will also prepare the next chapters, in particular Chapter 6 on contemporary money machines and Chapter 7 about material artefacts and places in contemporary finance.

First, there is the history of technologies for making coins, bank notes, and imprints, for recording transactions and for accounting, for making computer programs and screens that represent money by means of numbers on a screen. In Mesopotamia, clay tablets were used to record financial transactions (see also my citation of Ferguson earlier in this chapter). But there were also (other) artefacts that aided accounting. For instance, tokens were used to represent goods such as wheat and cattle. In Egypt, but also later elsewhere in the ancient world, papyrus was used for record-keeping and (tax) administration. Agricultural production, public works, labour, religious rituals – all these needed to be (centrally) managed. And then there is of course the production of money and imprints. In Lydia (in present-day Turkey) coins were produced with their value imprinted on them around 650 BC. Around 800 AD, the Chinese already produced paper money. Further down (or back) in the production chain are also mining technologies and other technologies that make whatever is needed to produce, for instance, linen and cotton paper, printing presses and electronic hardware. In all cases there are links to wider social relations. The relation between mining and slave labour has already been mentioned (in the ancient world, but we may also think of the historic silver mines that produced for the Spanish in Bolivia, in particular in Potosi). Even today there are clear links between, on the one hand, the production of precious metals, electronic technologies, and mining, and, on the other hand, exploitation, colonization, and war. Consider, for instance, violence and conflicts in central Africa.

Yet, as I suggested before, these technologies also have a more subtle moral and epistemic influence. There seems to be an increasing gap between, on the one hand, things that have direct, experiential value to us and the trade of which requires personal involvement, respect, ritual, and tradition, and, on the other hand, things that are only used as means and are entirely impersonal, e.g., means of exchange. When the cow became a coin, a distance was created that now seems hard to bridge. When money became dematerialized, new economic and political possibilities opened up, our world became larger and perhaps more interesting, but it also became a world where trust depends on what is written down and recorded, a world of impersonal trade relations.

Second, there are histories of machines that process numbers, calculate, assist in managing and processing financial information, technologies that are used in counting and accounting. In the history of money there have always calculating machines: calculators or indeed *computers*. There is history to be told of financial calculating machines, starting perhaps with the abacus and ending with computers, algorithms and computational finance.

The abacus or counting frame was used in Mesopotamia already in 2700–2300 BC in Sumeria, and probably also in Babylonia. Herodotus mentions that the ancient Egyptians also used it. Later the abacus was used also in Persia, India, China, ancient Greece, and the Roman Empire. In the Renaissance, Blaise Pascal invented a mechanical calculator. He created the machine in order to help his father deal with the tax affairs of Haute-Normandie. In the eighteenth century further calculating machines were invented and in the nineteenth century they were mass produced. In the twentieth century electronic calculators were developed in the wake of the first mainframe computers and in the 1960s miniaturisation of components accelerated and they became widely available.

Some were programmable. In the 1970s the first pocket calculators appeared. Personal computers were also first seen as calculating machines. When they could do word processing and had a graphic interface, that image changed. Today most electronic devices come with a calculator program/app. In the world of finance, computer models and algorithms are used in trade and exchange, for example, in high-frequency trading (see Chapter 6).

Again we can offer an interpretation of these technologies in terms of distancing. Does calculation imply distancing? Does it make relations to the world and to others more 'calculating'? It is again plausible that all these devices were and are *distancing technologies*. They did not only calculate; they also made humans and human society more 'calculating'. And this is not only true for 'calculation' technologies strictly speaking, but also for bookkeeping technologies and financial management systems. Let me explain this.

As I already suggested in Section 2.2., financial technologies and financial management made it possible for 'horizontal' financial and economic relations between people to be partly replaced by 'vertical' relations, which also had consequences for social relations, power, and responsibility. Financial technologies rendered it possible to *subject* people to new financial regimes and forms of power; they enabled and consolidated new vertical power relations. People were now under the control of the palace and the temple, and their bookkeepers. Goods were alienated from their context of production and consumption and were transferred to the central power.

This means that responsibility for financial transactions shifts from a 'horizontal' responsibility between trading people to a 'vertical' responsibility to pay taxes, to keep books for the palace, and do other things the new powers want. It means that not only the goods and the technologies, but also the control and the responsibility for trade is alienated from the people. They no longer respond to one another, but to the central ruler and the central law. Ethics and morality increasingly lose their personal dimension and their 'horizontal' social aspect and take on a 'vertical' mode: being a good person and doing the right thing is no longer something that is judged by one's fellow people (and indeed by oneself); it is judged by a higher power: by the one God and his servants, by the King, by the Law and (later) by Reason. It is no longer a personal and social matter; it is an *imperative*, that is, a command. There is an authority outside the person and outside direct social relations, for example, the *emperor* and/or the law, which dictates what to do. 'Vertical' principles and laws become prominent. What people do (wrong) is recorded and, if necessary, punished. The moral life becomes part of accounting. To be responsible becomes to be accountable. To do the right thing can only be decided by means of following a rule or by calculation. The moral life becomes a matter of *algorithms*.

Indeed, the financial technologies create conditions under which people are also made into calculating subjects and start to think about themselves in this way. In order to be able to subject people, they need to learn to read and write, and of course to calculate. They need to be able to read the laws that are prescribed to them. They are also disciplined into calculating and bookkeeping. First, only a small staff was able to do this, and they then further disciplined the people. But then in later times everyone uses writing and accounting technologies. Farmers had to keep books. People learned to read the laws. In modernity, an entire culture of accountability arises. Citizens and employees are disciplined; they have to make everything they do accountable. People in offices have to record what they do or they are recorded. Everyone has to fill in tax forms. Financial techniques and management systems are put in place to discipline employees. In contemporary bureaucracies such as corporations, government organizations, and universities, we have to *render ourselves accountable*. We have to keep records of what we do. We have to 'budget' our activities, say what exactly we are going to do, give specific objectives, and so on. We have to make our objectives and data transparent to the higher powers. There are still orders, commands, imperatives. But it is our responsibility to keep track, to keep records, to *discipline ourselves*. Similarly, we have to do our own moral bookkeeping and apply moral discipline to ourselves. Others record data and we record data. A system of moral self-surveillance is in place. The 'good life' (eudemonia) is not reached by means of self-reflection but by means of accountancy, data recording, and (self-)surveillance.

Yet 'horizontal' relations also become more calculating. They are no longer based on trust. The new vertical relations erode the horizontal ones. They even touch subjectivity. Calculation and accountability are internalized; subjectivity changes and so do relations between people. Instead of exercising direct responsibility to others – other traders, other workers, other members of the community – one is responsible to often *remote* higher powers: managers, civil servants, politicians. But, as suggested above, explicit commands are often no longer necessary; this remote, 'vertical' kind of responsibility is internalized. There is not only discipline but also self-discipline. We keep records, we produce data. In our times, we 'book face' (e.g., Facebook) in the sense that we are disciplined into not only keeping records of our goods and finances, but also of our personal activities, social relations, and private identity, which are commodified into data, self-managed and self-recorded, and then sold off to companies or transferred to the government. We have become currency. In the end we compute, calculate, and (re-)write ourselves. We become numbers in a 'book', data in a database, information in the cloud. We are both managed and 'pre-scribed' by others *and* managed by ourselves. Financial technologies and management techniques shape a calculating, managing subject.

This means that the moral geography of control and surveillance shifts from what Bentham and Foucault called a 'Panopticon' (Foucault 1975) in which disciplining is imposed from above and people (originally: prisoners) are kept under 'vertical surveillance', to a new form of surveillance, perhaps a new kind of 'Panopticon', in which people are far more active in rendering themselves visible and keeping themselves under surveillance. By means of social media, we participate in a 'horizontal' surveillance and control system and we discipline ourselves. They are part of what Foucault in his later work called 'technologies of the self' (Foucault 1982): practices – and I add, *material* technologies – by which subjects constitute themselves, work on themselves, govern themselves, discipline themselves. Moreover, whereas in the modern prison and modern society the prisoner, employee, and so on, is not held responsible and is disciplined by higher powers, the contemporary 'prisoner' is subjected, or rather subjects him or herself, in the name of freedom and responsibility. Self-management and self-commodification are presented as contributing to our autonomy; but our gain in external freedom (we are not *formally* slaves or prisoners and in the West there are fewer people who tell us what to do) has gone at the expense of a loss of internal freedom: our self and our person are now part of the moral-social-financial management system. Calculation and bookeeping has invaded the self.

Of course to some degree this has always been the case; the construction of the calculating subject, 'the number', started already in ancient times. By the time modernity sets in, the history of civilisation, here interpreted and understood as one shaped by writing/reading and calculation techniques, has produced two types of subjectivity which seem to be contradictory. On the one hand, it has produced a subject which understands itself as having an 'inner' and 'outer'. This is produced mainly through writing and reading, which makes possible a division between 'external' and 'internal' thought, between 'what is in my head' and 'what is on paper'. On the other hand, it is has also produced a subject which understands itself as a commodity and a number – an understanding which defies any distinction between 'inner' and 'outer', and which annihilates the self (or expands it infinitely). Indeed, financial technologies not only make possible distance from social relations, but also distance from the self. The self is either created as an inner realm on which we then can reflect, or it is monitored and managed – by others and by our … self. Mind and body are self-disciplined and exercised. Financial technologies make possible both subjecting and self-subjecting, disciplining and self-disciplining. These forms of disciplining, subjecting and subjectivity constitute forms of *distancing*. Financial technologies, together with other technologies, not only change our economic relations strictly speaking (if that makes sense at all), but our entire lives, our subjectivity, and our social relations. New financial technologies and new management techniques mean, each time again, new forms of subjectivity, new social relations, new lives, new societies and cultures, and new forms of distancing. These new forms of distancing are in turn connected to new, different, forms of responsibility and morality.

In the next chapter I will further develop my arguments about distancing.

# Chapter 3

# The pure tool that distances: Simmel's phenomenology of money and its relation to philosophy of technology

## 3.1. Introduction

In a project that studies and evaluates financial technologies, analysis and reflection on what is perhaps the most significant financial technology of all times is mandatory, and Simmel's work occupies a key position in this domain. In this chapter, I review Simmel's phenomenology of money and analysis of modernity in order to better understand the relation between financial technologies and various kinds of distancing. Here my discussion still has a historical dimension, but its context moves in the course of the chapter from Simmel's modern finance to contemporary, *electronic* financial technologies in a global context – thus preparing the ground for the next chapter.

I write 'phenomenology' of money since I read Simmel not only as a sociologist but also as a philosopher and in particular philosopher in the tradition of *phenomenology* and a *philosopher of technology*. Let me briefly clarify this. First, Simmel is a phenomenologist; yet in contrast to Husserl, he is not interested in "pure consciousness" but in 'our cultural relation to the world' (Lethonen and Pyyhtinen 2008). He studies experience and what conditions and structures our experience, but he believes that these must be found in the social. This implies that he theorizes how money shapes the social, cultural world and is shaped by it. According to Simmel, the phenomenon of money is thus very much linked to social relations and values (Lehtonen and Pyyhtinen 2008). In my interpretation of his work on money I will highlight the social-phenomenological significance of money. Second, when Simmel shows how money as medium and tool makes possible modern society, he also teaches us something about the relation between technologies and society. So far, Simmel has not sufficiently been recognized as a philosopher of technology. In my interpretation of Simmel, I will expand his approach to financial technologies and read his work on money as a contribution to philosophy of technology.

First I extract from Simmel's work three related but distinct clusters of claims about, and analysis of, money. I focus on his analysis of the nature of money and its role as a mediator which objectifies and de-personalizes, his conception of money as a 'pure tool' and his suggestion that there is a history of dematerialization of money, and his discussion of the social and moral implications of money in the context of modernity. Then I start exploring what this view of money and its relation to modern society may mean for contemporary financial technologies in a global context, and indeed for today's electronic media and ICTs in general. Interpreting Simmel as a philosopher of technology, I also make connections to Marx and to contemporary authors in this field (Borgmann and Dreyfus) who highlight the disengagement promoted by modern technologies, and offer some objections to Simmel – and hence to my initial application of Simmel to contemporary financial technologies. This includes a brief exploration of Marx's view of money and labour. Finally, I offer a short preview of the next chapters, which will further elaborate but also nuance some of the arguments and insights gained by this reading and 'updating' of Simmel.

### 3.2. Simmel's phenomenology of money: Distancing, dematerialization, and alienation

Simmel's analysis in *The Philosophy of Money* (1907) can be broken down into three components, which each contribute to the view – this is my summary and interpretation – that money makes possible distancing, dematerialization, and alienation: (1) the nature of money as a mediator and the distance to objects and people it creates in that role, (2) the development of money into the purest of tools and the distance to materiality and substance this creates, and (3) the social and moral implications of money in the context of modernity, which also can be interpreted as the creation of a number of distances. As I will show, in each case there is a dialectic between bridging and distancing: money mediates, links, and bridges, but at the same time it also creates distance. In the next section, I will suggest that this interpretation also works for contemporary information and communication technologies in general, which, at a higher stage of dematerialization and modernity, seem to be moving towards their 'purest' form as 'pure means'. (See also the next chapter.)

Simmel's view seems to be in line with other critics of modernity such as Marx, Weber, and Heidegger. I will say more about this below. Let me already emphasize now, however, that only the third part of his analysis and types of distancing have to do with modernity *per se*, for example, the distancing that relates to the modern division of labour (see also, for instance, Marx's analysis as referred to later in this chapter). The other types of distancing and alienation, by contrast, are directly related to what according to Simmel lies in the nature of money itself (in particular its function in and as the exchange relation) and its historical development towards 'the purest of tools'. This wider scope renders his analysis relevant not only to thinking about *modern* financial technologies and societies, but also to thinking about earlier financial technologies and indeed about money and the role of technology in our civilization and its future.

Note already that, in contrast to Marx[1] and most Marxians, Simmel does not focus on capitalism. Some may regret this (see, for example, Frisby's remarks in the older introduction to the English translation) but this 'neglect' also makes possible an account of alienation and objectification that is relevant to understanding and evaluating a broad range of – existing or imaginary – societies and cultures that use money and advanced financial technologies, including societies that are/were not capitalist at all, or that are/were perhaps capitalist in a different way. Simmel's analysis works at a level of abstraction that helps us to better understand financial technologies and perhaps even technologies in general. I will interpret Simmel as a philosopher of technology and media (and of 'civilization'), not as a philosopher of capitalism and only partly as a philosopher of modernity.

Let me further unpack the three components of what I take to be Simmel's view of money.

### *3.2.1. Money as medium places objects and persons at a distance*

Money mediates the exchange relation as it inserts itself between objects and people; it thus functions as a kind of bridge.[2] But Simmel points out that it thereby also alienates things from people. This is not due to the specific form of money, for example, coins or bank notes, but to the exchange relation itself. According to Simmel, money represents the exchange relation: in order to fulfil its function, money has to be impersonal, detached from every specific content. In the exchange relation, the value of objects becomes objectified. The object no longer has value for me as a person; its value becomes abstract. It no longer has 'subjective significance'; its value is relative and depends on the value of other things. This means that its value comes to lie entirely outside myself, far away from my desiring and valuing. I am confronted with an 'objective realm' (Simmel 1907, pp. 79–81). Personal value and (as we will see below) personal relationships are replaced by economic value and impersonal exchange. Fulfilling its function as a 'completely neutral' medium (p. 123), money is about 'the value of things without the things themselves' (p. 121).

---

1 I will say more about Marx later in this chapter.

2 Note that this is only one of the meanings of 'medium' and 'mediation'. In this sub-section I use the term as referring to something 'in between' that functions as a bridge or connector between two elements. A different meaning of 'medium' is 'environment' or 'surrounding' as in 'The fish swims in the water, has water as its medium'. That meaning of 'medium' fits better with other parts of Simmel's analysis (especially his analysis of modernity, e.g., the calculating character of modern times: money and calculation are then all around us and in this sense they are our medium) and indeed with what I said in the previous chapter about management and accounting: my point could be rephrased as saying that these are not separate spheres of activity but that they have become the environment or indeed 'medium' in which we live, think, and relate to each other (and to ourselves).

To function as a medium, money itself has to be something abstract. Simmel writes that money is 'interchangeability personified' (p. 124) and explains this as follows:

> If money itself were a specific object, it could never balance every single object or be the bridge between disparate objects. Money can enter adequately into the relations that form the continuity of the economy only because, as a concrete value, it is nothing but the relation between economic values themselves, embodied in a tangible substance. (Simmel 1907, p. 125)

For Simmel, this relational character of money expresses not only the nature of the exchange relation and economic value, but also reveals that relational nature of all being. Here the force of Simmel's phenomenological approach becomes clear: money is not so much a thing that belongs to the sphere of 'economy', but a medium, a relation, and even – to use Heideggerian language – a way of revealing. It shows, reveals, and opens up the relationality of all being. As a very special kind of medium, it discloses the relationality of the world:

> The philosophical significance of money is that it represents within the practical world the most certain image and the clearest embodiment of the formula of all being according to which things receive their meaning through each other, and have their being determined by their mutual relations. (Simmel 1907, pp. 128–129)

Money thus shapes our experience and knowledge of the world by creating distance *and* by mediating relations – thereby even embodying exchange and relationality themselves. Both functions are related: to play its role as a bridge between objects and persons, money has to be different and distant from them. Money thus bridges and creates distance at the same time: it bridges and creates distances between persons and things and (as we will see below) also between persons.

Note that Simmel's focus on money as a medium of exchange may be criticized by those who connect money's origin and meaning in debt instead of exchange (see my remark in the previous chapter). But regardless of the precise origins of money, it is clear that in modern times the exchange function of money has come to be seen as a crucial one, one which is central to daily experience of money and which raises normative concerns about distance between people. One could even say that money as exchange medium has become the symbol of impersonal relationships in modern society. What renders the myth about the barter origin of money so powerful is perhaps this modern longing for more personal and more direct relations. Myths tell us a lot about our views, values, and indeed problems. Simmel's analysis, however, shows that the distancing problem is not limited to modern society and modern money, but has always accompanied money when and in so far as it functioned as a mediator. (And we may add: *will* always accompany money when and in so far as it functions as mediator.) Therefore, Simmel's insights into the relational, abstract, and distancing aspects of money as medium remain highly relevant, whatever its historical origins may be and whatever other functions and features it may have.

### *3.2.2. Money as a pure tool dematerializes our culture and creates distance between abstract thinking and material substance*

In order to play its role as neutral medium, does money still have to be substantial at all? Simmel argues that this is not necessary. In essence it is a pure symbol and function, not dependent on substance. However, for Simmel this symbolic and abstract character of money is not a mere coincidence. He argues that money represents a stage in the increasing intellectualization and spiritualization of culture; it is part of how our culture dematerializes. It is more an idea than an object. Moreover, Simmel argues that money as function is a 'pure means' and 'the purest tool'. Let me unpack both (related) views.

First, according to Simmel there is 'a profound cultural trend' (p. 148) towards the symbolic, the abstract, and the quantitative, and money is part of that trend in so far as its functional value 'exceeds its value as a substance', which in turn depends on the diversity of the services it performs and the speed of circulation (p. 144). As function, it becomes an idea, a symbol, a representation of 'pure quantity in a numerical form' (p. 150). This means money, and the culture it is part of, becomes less substantial, dematerialized. Simmel describes the final outcome of this trend as one in which only the function is left; money then becomes detached completely from every substantial value (p. 165). Although in his time it was 'not technically feasible to accomplish what is conceptually correct' (p. 165), he sees paper money as signifying 'the progressive dissolution of money value into purely functional value' (p. 172).

This functional value, however, is not only related to distancing from objects (bridging and creating a distance between persons and objects), but also to distancing from persons (bridging and creating a distance between persons). Again this is due to the nature of money, which according to Simmel is a *social* one (see also the next component of his view). In Simmel's view, money is not only an economic phenomenon but mainly or even 'entirely a sociological phenomenon' (p. 172). Money refers to the activity of exchange, which is a relation between people. He says that 'the relations between the objects are really relations between people ... money is the reification of exchange among people' (p. 176). And if money is a social institution (p. 209), this implies that the history of money is also the history of social relations. The time that money would embody pure function is also the time when social relations would have reached an advanced stage (p. 172). Simmel envisages an 'ideal social order' in which 'money with no intrinsic value would be the best means of exchange' (p. 191). This would be the end point of a process Simmel calls 'the growing spiritualization of money' (p. 198): once we realize that money is 'a pure symbol' (p. 202), we no longer need a material basis. (In the next chapters I will ask if we have reached that stage.)

But money is not only a matter of social relations; it is also a tool, and according to Simmel it is a very special one. He refers to Adam Smith, who wrote in *An Inquiry into the Nature and Causes of the Wealth of Nations* that gold and silver are like kitchen utensils (p. 173). But we have seen that it is a tool that is independent of its particular substance, if such substance is necessary at all, and this gives it particular properties. Let us further look into Simmel's conception of money as a tool, which will deliver further insights into money as medium and technology.

Tools are always media in the sense that they are 'in between' other elements, their use 'involves interposing an intermediate position' (p. 209). As purposeful beings, we introduce intermediary steps between ourselves and our ends. Man is therefore not only a purpose-oriented being but also a 'tool-making' animal. Tools are instruments that insert themselves in teleological chains, in the sequences of means and ends. Tools are a kind of means (Simmel says an 'intensified' means). But money is a special kind of tool since it is 'a tool of endlessly diverse and extensive uses' (p. 210). Therefore, Simmel calls money 'the purest example of the tool' (p. 210): it has no purpose of its own (p. 211) but is a pure means. It is in this sense also a pure or 'perfect' tool:

> a tool continues to exist apart from its particular application and is capable of a variety of other uses that cannot be foreseen. ... Money as the means *par excellence* fulfils this condition perfectly; ... Money is the tool that has the greatest possible number of unpredictable uses. (Simmel 1907, p. 212)

Money is not connected to a specific end at all; which end it will be used for remains open. Simmel calls this 'the pure potentiality of money as a means' (p. 218). Money is highly flexible in this sense, it adapts to every purpose: 'money's flexibility is only that of an extremely liquid body which takes on any form, and does not shape itself but receives any form it may possess only from the surrounding body' (p. 329).

However, at the same time, exactly because it is the purest and most perfect tool, it becomes the final end; if money can do everything, then it comes to be valued in itself: the means becomes an end (p. 232) and develops into an independent value (p. 234). If we believe that it is omnipotent (p. 237), it becomes the final purpose. It becomes a power that, in a very monotheist fashion, tolerates no other end next to it. In the money economy, 'specific values' fade into 'mere mediating value' (p. 257), whereas money becomes the ultimate value. This leads us to Simmel's analysis of the social and moral consequences of money, especially in the context of modernity and globalization.

### *3.2.3. Money under modern conditions bridges and creates distance*

When Simmel discusses the social nature of money, he already suggests that what we would now call 'globalization' has important social consequences and is closely linked to the development of money. Money has always made possible an extension of the economic space (which in turn required money to bridge the increased distances), but this also means both an enlargement of our social environment and a reduction of direct contact between people (p. 181). On the one hand, the social sphere grows. Simmel notes a close relationship 'between the money economy, individualization and enlargement of the circle of social relationships' (p. 347). Money, as 'the most mobile of all goods' (p. 354), 'forms the bond that combines the largest extension of the economic sphere with the growing independence of persons' (p. 349). We become closer to those remote from us. On the other hand, at the same time we become more remote from those close to us. Simmel describes this social-spatial process as follows, using the language of distance and distancing:

> Modern man's relationship to his environment usually develops in such a way that he becomes more removed from the groups closest to him in order to come closer to those more remote from him. The growing dissolution of family ties; the feeling of unbearable closeness when confined to the most intimate group, in which loyalty is often just as tragic as liberation; the increasing emphasis upon individuality which cuts itself off most sharply from the immediate environment – this whole process of distancing goes hand in hand with the formation of relationships with what is most remote, with being interested in what lies far away, with intellectual affinity with groups whose relationships replace all spatial proximity. The overall picture that this presents surely signifies a growing distance in genuine inner relationships and a declining distance in more external ones. ... Thus, correspondingly, the most remote comes closer at the price of increasing the distance to what was originally nearer. (Simmel 1907, p. 576)

Money and (what we today call) globalization are thus closely linked. However, the reason why money can play this role is again connected to the nature of money. In Simmel's view money contributes to, and is even constitutive of, the individualization and disruption of social relations in modernity for the very same reasons as it was always already socially alienating – albeit perhaps to a lesser extent: 'exclusively determined by quantity ... money is completely cut off from the corresponding relationships that concern it' (p. 279). In modernity, this effect increases significantly in combination with the modern division of labour, which 'causes personalities to disappear behind their functions' (p. 296). If persons are reduced to their function in the labour process and to their monetary exchange value and cost, and if internal and external relations mainly take place through the medium of money, then a distance is created between one's personality and the work activity, and between persons. Simmel thought that what we produce should be 'a reflection of the total personality' (p. 454), similar to the creation of a work of art or to the work of the medieval craftsman. But he argued that this is rendered impossible if 'the worker places his labour at the disposal of another person for a market price and thus separates himself from his labour' (p. 456). Interestingly, this happens to the worker but also to the manager, who is equally subject(ed) to an 'objective' purpose and 'now produces *for the market*, that is for totally unknown and indifferent consumers who deal with him only through the medium of money' (p. 335). Whereas the custom work done by the medieval craftsman involved 'a personal relationship to the commodity' (p. 457) and indeed between craftsman and customer, in the modern economy there are only impersonal relationships with consumers and products. Money, together with modern machinery, produces indifferences towards objects. Simmel writes:

> the sheer quantity of specifically formed objects makes a close and, as it were, personal relationship to each of them more difficult ... What is distressing is that we are basically indifferent to those numerous objects that swarm around us, and this is for reasons specific to a money economy: their impersonal origin and easy replaceability. ... Objects and people have become separated from one another. (Simmel 1907, p. 460)

Compare a personal gift from someone you know (e.g., your child, spouse, neighbour, etc.) that has been made by that person with an object you buy in a store: the latter is impersonal in origin and easily replaceable, whereas the former is not. This impersonal character of objects also means that the (new) owner of the object has an impersonal relation to it. Money detaches objects, they lose their personal significance. Objects and possessions become disconnected from persons and their activities: 'To the peasant, the land meant something altogether different from a mere property value; for him it meant the possibility of useful activity' (p. 399). Since money 'intervenes between person and commodity', we lack the immediacy of things and 'are no longer directly confronted with the objects of economic transactions' (p. 478). Money creates 'spatial distance between the individual and his possession', which has consequences for personal engagement and indeed personality and morality:

> The owner of shares who has absolutely nothing to do with the management of the company; the creditor of a state who never visits the indebted country ... this is only possible by means of money. Only if the profit of an enterprise takes a form that can be easily transferred to any other place does it guarantee ... a high degree of independence ... The power of money to bridge distances enables the owner and his possessions to exist so far apart that each of them may follow their own precepts to a greater extent than in the period when the owner and his possessions still stood in a direct mutual relationship, when every economic engagement was also a personal one ... Because of this separation of object and person, the ages of highly developed and independent technology are also the epochs of the most individualistic and subjective personalities. (Simmel 1907, p. 333)

According to Simmel, therefore, there are social and moral consequences. Money makes possible not only indifference towards objects, but also indifference towards people. There is a distance between owners and their possessions, but also between salespersons and consumers (as previously indicated) and between people doing business. Money creates social and moral distance. It makes possible new forms of sociality and new ways of producing and doing business. Simmel writes that money has brought about more individualist forms of social association, and refers to 'the joint stock company whose shareholders are united solely in their interest in the dividends, to such an extent that they do not even care what the company produces' (p. 343). Thus, not only workers, but also consumers, managers, and owners are alienated from the production process, from the product, and from each other. There is what Marx called 'indifference' (Marx and Engels 1846, p. 66) and loss of 'self-activity' (p. 87). There is a lack of personal engagement with things and people. What remains is an objective purpose: the goal of 'making money'.

Simmel remarks that for this purpose one does not need a lot of skills. He writes that 'a large number of occupations in modern cities' require only very general and abstract skills and he identifies bankers as 'the class with the most general and abstract functions' (p. 437). Again this follows from his analysis of money: if money is the symbol of abstraction, then financial occupations are the most 'abstract' ones. They are also among the most 'calculating' ones. Money promotes 'the calculating character of modern times' (p. 443):

> The money economy enforces the necessity of continuous mathematical operations in our daily transactions. The lives of many people are absorbed by such evaluating, weighing, calculating and reducing of qualitative values to quantitative ones. (Simmel 1907, p. 444)

In the end, people also become mere instruments as only rational and objective relationships are promoted in the money economy. Simmel compares the nature of money with 'the essence of prostitution': the problem is 'the objectivity inherent in money as a mere means which excludes any emotional relationship', which leads to 'mutual degradation to a mere means' (p. 377). (Note again the mutuality of the personal, social, and moral degradation: objectification happens not only to the workers, but also to the capitalists, not only to the producers but also to the consumers, etc.)

Simmel mentions many more problems. For instance, he argues that money also makes possible remoteness from nature: the distancing in the economic realm also makes possible 'the distinctive aesthetic and romantic experience of nature': Simmel suggests that the distance from nature 'is the basis for aesthetic contemplation' (p. 479). Indeed, he even says that 'all art brings about a distancing from the immediacy of things' (p. 473). Furthermore, because of its anonymous and colourless nature, money also makes possible illegal activities: 'it does not reveal the source from which it came to the present owner' (p. 385). And Simmel warns that we are becoming the slaves of our products and our technologies (p. 483). He complains that, to put it in McLuhan's words, the medium has become the message:

> It is true that we now have acetylene and electrical light instead of oil lamps; but the enthusiasm for the progress achieved in lighting makes us sometimes forget that the essential thing is not lighting itself but what becomes more fully visible. People's ecstasy concerning the triumphs of the telegraph and telephone often makes them overlook the fact that what really matters is the value of what one has to say. (Simmel 1907, p. 482)

This distance between medium and message means that, according to Simmel, we become estranged from ourselves, from our 'most distinctive and essential being' (p. 484).

### 3.3. Updating Simmel, or the present and future of money: Electronic monies and currencies as pure tools and means of social and moral distancing

There are a number of objections one can make against Simmel's analysis of money, technology, and modernity (see the next section), but let me first show how relevant his analysis still is for understanding contemporary financial technologies, in particular forms of electronic money and currency. I will continue and further develop this analysis in the next chapters, but on the basis of what Simmel says let me offer the following preliminary observations.

First, contemporary forms of money are made possible by electronic information and communication technologies (ICTs) which seem to create an 'objective realm' of financial numbers and monetary information – in the form of 'data' – which are removed from personal relations with objects and between people. The numbers on the screens of traders express values and relations, but they do so in an entirely abstract, impersonal, and disembodied way. There is no immediate link to the goods and the people involved. For instance, traders at the stock exchange, sitting behind the screens of their electronic trading platforms, have no immediate link to the goods they are trading and to the people they are dealing with – let alone to the people that are affected by their activities. (See also Chapter 6.) The most important people in this kind of trading are the so-called 'quants': mathematics experts who use computers to analyse and trade on the market. It is a world of numbers, a world in which machines are more at home than humans. (Indeed often machines take over; see Chapter 6.)

Second, the specific form electronic money takes is removed from all 'tangible substance'. Electronic money, and especially electronic currencies such as Bitcoin (see Chapter 5), seem to represent a late if not the final stage in Simmel's history of money: they seem to be located at the point where money becomes a pure symbol and function, not dependent on substance at all. Numbers on a screen have no substance, bits of information have no substance. There is only 'pure quantity in numerical form' (Simmel 1907, p. 150). With a currency such as Bitcoin, money is detached from any substance, it is pure information, pure data, pure quantity: today it is 'technically feasible to accomplish what is conceptually correct' (p. 165). Electronic money and currency thus seems to illustrate the financial dimension of the spiritualization of culture. It looks as if 'the progressive dissolution of money value into purely functional value' (p. 172) has taken place. Furthermore, electronic money fully reveals the functional and sociological value of money, and, more generally, the relationality of being. Currencies such as Bitcoin, which do not rely on support from a centralized bank or on a link to precious metals, depend entirely on trust of people; it is a matter of (exchange) relations – not substance. Bitcoin is a decentralized currency, and bits have no intrinsic value. Bitcoins are generated by a computer algorithm, and again the numbers on a screen are pure symbols, there is no material basis. We can now dispense with the old kitchen utensils in the house of global finance. It seems that with the use of electronic forms of money we have reached Simmel's 'ideal social order', a utopian (or if you like: dystopian) financial and social world.[3]

Third, electronic technologies – including those used in finance – seem indeed 'the purest example of the tool' (p. 210): they are pure means or perfect tools in the sense that they are extremely versatile. Electronic devices, in particular, are capable of many uses, many which cannot be foreseen. Compared with other tools, they seem to have 'the greatest possible number of unpredictable uses' (p. 212). They do not have a specific end attached to them; the end is open and depends entirely on the user. They are highly flexible and, provided there are enough software programmes or apps available, they very much resemble the functional flexibility of 'an extremely liquid body which takes on any form, and does not shape itself but receives any form it may possess only from the surrounding body' (p. 329). Both money and data are said to exhibit a great degree of liquidity[4]. Combined with the right hardware they seem to be capable of flowing into any desired means; they seem to be universal tools.

Fourth, because electronic devices display some degree of omnipotence, the possession of them becomes for some people an end in itself, and sometimes the desire to have them is even a motive to use violence. They thus resemble money also in this respect.

---

3 Indeed, as Dodd has argued, Simmel's arguments about 'perfect' money can be connected to his idea of a perfect society, his views on money and socialism, and the relevance of his ideas for thinking about how to transform and improve society (Dodd 2012; Dodd 2014). (I will return to this in the last chapter.) But the utopian dimension of this philosophy of money can also be interpreted as referring to the link between (perfect) money and globalization: 'u-topia' means 'non-place'. I will say more about this below.

4 Note that the term 'liquidity' is used here in the common metaphorical sense, not in the technical sense used by economists and finance experts.

Note that, of course, in practice it is not quite true that 'there is an app for everything', and 'algorithm for everything' (compare with 'money can buy everything'). But the Simmelian point here is that there is a historical development towards more liquidity, towards an all-purpose tool. Furthermore, with Simmel we can also explain why they are not fully omnipotent: electronic devices themselves are still material. They are not 'perfect tools' yet. Software is highly flexible and versatile, but the hardware remains material. Money, by contrast, mediates our purposes – to quote Simmel again – 'in the same manner as other technical mediating elements, but does it more purely and completely' (p. 485).

Fifth, an electronic currency such as Bitcoin is often thought of as anonymous and colourless, which, if true, makes it also the 'perfect' money for illegal activities: as Simmel says, money 'does not reveal the source from which it came to the present owner' (p. 385), and it seems that Bitcoin performs this function very well since it has been used for activities such as drugs and arms trade and gambling. (However, Bitcoin is only anonymous at first sight. Electronic money is more traceable than cash, and it turns out that this is also the case with Bitcoin. In fact, every transaction is publicly logged and it is in principle possible to trace the identity of someone if any of the addresses in the transaction history can be linked to a person's identity. Work has to be done to increase anonymity. Cash is far more anonymous.)

Sixth, both electronic technologies and new forms of money are very much linked to the phenomenon of globalization. These new technologies helped and help to extend the economic space to the globe; now there are global streams of electronic money that never stop. Physical distances are bridged all the time, and global financial and social interdependence increases. However, as Simmel suggested, this enlargement of the circle of relationships and mobility of money comes at the price of 'increasing the distance to what was originally nearer' (p. 576) – or at least so it seems. It seems that electronic money contributes to the decreasing importance of place, of the 'local': the global world of electronic money and media is also utopian in the sense of being a non-place. Electronic money is everywhere and nowhere, and so are we, it seems (in Chapter 7 I will criticize this).[5] Moreover, both electronic devices and their users have become more mobile. The mobile phone becomes a symbol of global mobility as such, of utopia as every-place and non-place, the abolishment, absolute shrinking of

5 There are also other meanings that can be attached to utopia as 'non-place'. For instance, utopian thinking may be interpreted as wanting to abolish distance between persons, desiring an absolute immediacy (some may expect ICT to create this, perhaps even at global scale, or others may instead fear that technology makes this immediacy impossible), or creating an unbridgeable, infinite distance between present and future in so far as it projects an ideal, perfect world which can never be reached. It can also be understood as creating infinite distance in the sense that the thinking is entirely abstract, at infinite distance from life 'on the ground', from concrete practices and contexts.

distance.[6] Information is extremely mobile and is also a medium[7] which becomes 'meta', very much in the same way as Simmel viewed money: to fulfil its function as universal, perfectly liquid medium it has to be removed from the specific relations between people and between people and objects, removed from direct experience. What is left are data, which are sold and processed as a commodity. They come to be valued as such. We want information, and we forget why.

Seventh, the global electronic market, together with division of labour, has created further impersonal relationships to both goods and (other) persons. Simmel's remark about 'the owner of shares who has absolutely nothing to do with the management of the company' and the 'creditor of a state who never visits the indebted country' (p. 333) is still very relevant to today's global economic and financial system. It seems that globalization has only produced more distance: distance between persons and goods, distance between persons. People who make investments at one end of the world do not really know what happens at the other end of the world, where their money is put to work. This is due to the distance made possible by electronic money, and also to complex financial products, which, as highly abstract things (one may say 'ideas'), further alienate the trader and the investor from the objects of investment and the (other) people involved. Such financial products are also made possible by electronic technologies.

This distance to people and things has moral consequences. Indifferent shareholders want to make profit and 'do not even care what the company produces' (p. 343). Often a contemporary company produces nothing (in the sense of 'no-thing'): it only deals with money, with numbers, with symbols far removed from the world of people and from the material world. Financial occupations are indeed among the most 'abstract' occupations we know. As indicated before, people in the financial world who are extremely good in it are called 'quants' and are engaged in mathematics and information science: they try to find better algorithms that predict the stock exchange and that take over trade decisions. In so-called 'high-frequency trading', trade is automated and takes place at a very high speed – a speed that can only be followed and reached by machines. These people and their technologies can be interpreted as the summit of the quantification, de-personalization, and fast rhythm Simmel described when he analysed modern life and modern society.

Eighth, since the rise of the internet and related technologies there has been a criticism that mirrors Simmel's remark about 'the triumphs of the telegraph and telephone': the essential thing is not the internet, the Web, social media, and all kinds of electronic devices, but the content and (I would add) what becomes visible in and through these technologies. What matters is not the progress in communication, but what one has to say. There is a widespread feeling that while the internet has brought us many benefits, it has also come at a price: a loss of immediacy, engagement, personal contact, disorientation or loss of a sense of

6 A different question is if it is a *good* place, if a non-place can be a good place at all.

7 Here medium means 'in between' but also 'environment': we live within an informational environment, an informational sphere.

place. Electronic money is a great tool, but it has also removed us from local goods and people, or so it seems. Indeed, the Simmelian point is not only that 'remote' relations are more superficial, but that our *local* relations are also influenced by the new technologies and media, such that even the proximate becomes more distant.

Finally, in the end it turns out that as 'intermediate' beings who need, develop, and use tools, we become tools ourselves, intermediate means rather than ends: we lose specific skills and close relations to specific others and specific places. We can be used everywhere. The age of electronic technologies is the age of the manager and the age of mobility: mobile work and mobile living. As information workers, indeed information processors, we need only general skills. We become ourselves all-purpose tools in the service of what Simmel calls 'objective' purpose. This also involves a levelling of the positions of worker and manager. As electronic tools, we are all workers and managers at the same time. What matters is the speed of information processing. Data must flow, money must flow. There is (still) a high degree of division of labour. Yet if, and to the extent that we are becoming more perfect means ourselves, we come to be seen as valuable and useful everywhere. To the extent that we become an impersonal tool, a medium, we become currency, we come to embody the exchange relation itself. We are supposed to work everywhere with everyone doing everything. Becoming all-purpose decreases our personal embodied value but increases our exchange value. We can take on any form, we are like liquid. We are liquid, as liquid as money. Information flows, we flow. Welcome to 'liquid modernity' (to re-use Bauman's term).

Again, objections can be made to what is likely to be read as at least an exaggeration of a tendency in contemporary society. It may be that in reality things do not look as bad as presented here. I hope that is the case. But at the same time, Simmel's work gives us a coherent and hermeneutically powerful tool to articulate what may be problematic about the use of the new financial tools.

This application and interpretation of Simmel's analysis thus warns us for all forms of distancing brought about by new electronic technologies, including new forms of money and currency. But Simmel did not think the development towards pure tools is bad in all respects. On the one hand, these new ICTs and forms of money seem to represent Simmel's utopian ideal of the spiritualization of human culture through – among other things – the development of the 'purest' of tools. On the other hand, with Simmel we also see the darker sides of this history; in particular we see that the bridging of many (physical) distances also implies the creation of new, moral and social distances. Our tools have moral, social, and anthropological consequences. And whatever these consequences are, this Simmelian analysis shows again that there is a close relation between financial technologies (e.g., electronic ones) and developments in wider society and culture. Perhaps some of Simmel's arguments can be extended to technologies in general. In the next section, I propose to read Simmel as a philosopher of technology. I will also briefly compare him to other philosophers of technology, and offer some objections to his account of money and what I take to be his (usually implicit) account of modern technology.

## 3.4. Simmel's contribution to philosophy of technology and some problems

### *3.4.1. Simmel's contribution to philosophy of technology*

Simmel helps us to better understand and evaluate financial technologies and electronic technologies, but also technologies in general. He makes a contribution to philosophy of technology in at least the following ways:

First, Simmel's analysis is in line with other critics of modern technology such as Marx, Weber, and Heidegger, who also connected technology to modern society and culture and its problems. Technology turns out to contribute to various forms of modern alienation, objectification, and disengagement. For instance, in the *Economic and Philosophic Manuscripts* of 1844, Marx distinguished between various forms of alienation (*Entfremdung*) in the labour process under capitalist conditions: there is alienation of the worker from the product of his labour, from working, from his species-essence, and from other workers (Marx 1844a). Marx writes: 'Labour produces not only commodities: it produces itself and the worker as a commodity' (p. 71) and alienation means here that 'the worker is related to the product of his labour as to an alien object' (p. 71). Moreover, labour itself is alienated: it becomes 'external to the worker' in the sense that it does not belong to him and that he 'does not develop freely his physical and mental energy but mortifies his body and ruins his mind. The worker therefore only feels himself outside his work, and in his work feels outside himself. ... His labour is therefore not voluntary, but coerced; it is forced labour' (p. 74). The labour does not belong to the worker. Furthermore, whereas man's 'species-being' (*Gattungswesen*) implies that in contrast to animals 'man freely confronts his product', in alienated labour free activity is replaced by an estranged life in which his species life becomes a means (p. 77). Finally, in this labour process humans become estranged from others: 'each man views the other in accordance with the standard and the position in which he finds himself as a worker' (p. 79). Under capitalist conditions there is thus a four-fold distancing in the labour process: between the worker and his product, between the worker and labour, between the worker and his species-being, and between the worker and others.

In this view, technology is an instrument of alienation and exploitation. Technologies, as means of production, are controlled by capitalists and create an industrial system of production which makes the workers into instruments, mere objects. Increased specialization creates less skilled and less enthusiast workers. They become part of the machine. This criticism of technology is in tune with that of other critics of modernity. Weber and Heidegger have argued that modern science, with its specialization and calculation, has resulted in disenchantment (Weber) and disengagement (Heideggerian philosophers – see below). The scientific world-view and the mediation of modern technology has resulted in a secularized view and a detached attitude and disengaged life. The world only appears to us as what Heidegger calls a 'standing-reserve' (Heidegger 1977), something for us to study and to control and exploit. Modern science and technology make it possible for us to be alienated and disengaged from things, from nature, and from others.

Today the disengagement argument has also been made by philosophers of technology, especially philosophers influenced by the phenomenological tradition such as Dreyfus and Borgmann. For example, Borgmann has argued that technological devices, by making life easy for us and retreating to the background, have made us engage less with things and people (Borgmann 1984). And Dreyfus has argued that our 'obsession' with the internet leads to disembodied experience and endangers the possibility of commitment (Dreyfus 2001) and that in contrast to the craftsman, who achieved 'intimacy' with material and had a 'feeling of care and respect for it' (Dreyfus and Kelly 2011, p. 210), modern technology makes us lose our sensitivities and threatens the possibility of meaningful, skilled engagement with the world. Contemporary interpretations of Simmel in relation to new media and technologies can make a significant and rich contribution to this tradition of thinking.

Second, Simmel also makes a contribution to philosophy of media and technology which goes beyond criticism of *modernity* as such: he suggests that moral and social distancing is not only a modern phenomenon but is deeply rooted in the material-cultural development of civilization, in particular in the tools we use and their connection to human relationships and activities. Simmel writes about money, but part of what he has to say concerns the human use of tools in general, in particular their 'sociological' character and teleological and therefore anthropological significance: as purpose-oriented beings, we introduce tools in the sequence of purposes, we are 'intermediate' beings. Very much in line with other philosophers of technology, Simmel notes that our tools always have unintended effects: 'a tool continues to exist apart from its particular application and is capable of a variety of other uses that cannot be foreseen' (Simmel 1907, p. 212). This is not a remark about modernity, it is a remark about technology and its wider impact. Similarly, contemporary philosophy of technology teaches that technologies – even non-modern ones – are never 'mere means'; they always do more than being instrumental towards the aim for which they are designed. They influence and shape our lives and our societies. Simmel focuses on money as a perfect tool and a pure means, which becomes entirely the end. But this also means, I submit, that other, non-perfect tools will always at least *influence* our ends, that they are never mere means. This interpretation of Simmel opens up the way for interpreting a variety of tools – also tools used before modernity – in terms of their distancing (and other) effects. Recall for instance the financial technologies mentioned in the previous chapter: the tablets, calculation devices, and the writing used for book-keeping and management had effects beyond their original purpose. Thousands of years ago, in the cradle of our civilization, these tools already contributed to a more 'objective', 'calculating' view of the world and indeed a more 'objective' and 'calculating' treatment of other people. Simmel rightly understood that money is about human relations and human society, and suggested that this is true for all tools, for all media and technologies. Extending

Simmel's view of money to all technologies, we can say that technologies are about human relations and human society. They are not stand-alone instruments but concern work relations, exchange relations, political relations. They help to shape our lives and our society.

Third, it seems to be Simmel's distinctive contribution to draw our attention to the distancing effects of a multi-purpose tool, especially an all-purpose tool. His analysis of money helps us to understand why exactly computers may disengage us from people and things: the problem is not only, and not so much, that they work with abstract symbols, numbers, and writing which are removed from what really matters (also literally: from materiality); the main problem – which is at the same time the explanation of their greatest 'triumph' – is that they can be used for (nearly) every end and therefore tend to absorb all (other) ends. This gives us a fresh perspective on contemporary computing and, more generally, contemporary media and ICTs. In so far as electronic ICTs can be used for nearly every purpose (this is at least their 'promise'), they may also absorb ends and become an end in themselves.

Furthermore, Simmel also makes his argument from in terms of distraction or even bewitchment, which can again be applied more widely and in a contemporary context. We are bewitched by the medium and come to think that information and communication as such is the only thing that counts, that 'being connected' as such is more important than what is being said and what is being exchanged. In finance, we think that the numbers on the screen are more important than the values and concrete relations they refer to.

This view is not without problems. For instance, both Simmel's formulation in terms of an 'objective' versus a 'subjective' world and this interpretation and application to contemporary ICTs still presume that there is a separation between 'medium' and 'what is really valuable', between means and ends. But we can take Simmel as a starting point to question this separation. For instance, perhaps the most urgent moral and social problem with computers, electronic devices and the internet is not that they distract us from relations (to humans and to objects) and from what is really valuable but, rather, that they have absorbed and transformed those very values, objects, and relations into information, data, numbers, bits. If the medium really becomes the message – to use McLuhan's phrase again – then there is only the medium, then all values collapse into the medium. Then, ultimately, the distance between the 'objective' and the 'subjective' also vanishes, implodes. Then the subjective can only be accessed in and through the objective, then the real can only be accessed in and as information. Is this a problem, or a solution? I return to this issue below when I articulate and discuss some possible objections to Simmel and the Simmelian interpretation of contemporary (financial) ICTs offered here.

Furthermore, as already noted in practice ICTs are not such a 'perfect' means and do not and cannot fully play this role of a medium as a kind of 'black hole' in which all value and meaning disappears (and reappears as information, as data). But even as 'imperfect' means which cannot be used for *all* ends they seem to create distance: they replace the immediacy of personal, embodied experience with a different kind of experience, in which all being is objectified, turned into data.

Note also that Simmel's philosophical anthropology is in line with Plessner's recognition that humans are naturally artificial (Plessner 1928) and with contemporary philosophers of technology, such as Verbeek and De Mul, who are inspired by Plessner. Simmel's critical philosophy of money is thus not directed against technology as such; on the contrary, as I have pointed out he recognizes that we are tool-making animals and 'intermediate' beings. We need technology and media as purpose-oriented beings. We also need it to develop ourselves (and here it seems that the objective and subjective meet). In 'The Crisis of Culture' Simmel writes: 'For culture, after all, means the cultivation of the individual by means of the cultivation of the world of things' (Simmel 1997, p. 92). Thus, Simmel connects the cultivation of culture with the cultivation of things. This renders him once again interesting and relevant as a *philosopher of technology*, and not merely as a philosopher of money (which is how he is usually read).

However, as I already indicated there are a number of problems raised by this interpretation and application of Simmel. Next to the previously mentioned controversial assumption that money originates in barter (which is, in the end, an empirical question) there are a number of *philosophical* assumptions and claims that deserve our attention. Let me present a selection of potential objections.

### *3.4.2. Some problems*

The first problem I wish to discuss connects directly to Simmel's characterization of modernity as involving a distancing between an *objective* world of things and technology and a *subjective* world of experience and value. According to Simmel there is gap between the two – partly because of money – and in modernity it widens. This view raises at least two problems. First, is this distinction tenable, given that tools are so much connected to the teleological chains, which according to Simmel are very human indeed? If we need the world of things in order to reach our purposes and if we need things to become the persons and subjects we are, then is this world of things not also partly subjective given their influence on our subjectivity, and is the 'subjective' world of human experience and values not always mediated and shaped by the tools we use, thus having a (not necessarily problematic) 'objective' aspect? Second, even if the distinction holds, then is it true that with money and in modernity the gap between 'objective' and 'subjective' widens? If the nature of money is 'sociological', as Simmel says, and if it depends on trust and human relations, then why would it render the world more 'objective'? It may distance us from others and from things, but the money economy itself seems to be 'objective' and 'subjective' at the same time. And does the gap really widen in modernity? Perhaps it did, at first, but information technology seems to change things. If and to the extent that, in contrast to the industrial age, information technology is a more versatile, nearly all-purpose tool, it seems to include more of the 'subjective' than ever before. It seems that when they become 'perfect tools', ICTs draw in all subjectivity and suck in all human values, perhaps transforming them. But that does not mean that they are gone. Is it necessarily bad if means and ends merge? Let me explain this further.

Computers may have been developed for the purpose of calculation, but as we can also say from a Simmelian perspective, they have and had many unintended consequences. As the history of computing unfolded, they were no longer seen as mere 'calculating machines' (and hence as 'objective' in Simmel's terminology) but became more entangled with (other dimensions of) human experience, feeling, and social relations. They came to be seen as more 'human' and 'social'. Today, with all the different applications and devices which merge and mix with our lifeworld, the original 'objective' purpose (for instance calculation) seems to be transformed into different, more 'subjective' purposes (for instance shaping one's identity and connecting with others). The 'non-human' calculator has become a 'human' device. Electronic equipment, both hardware and software, promises to do a lot more than calculation; today we use them for all aspects of life and for all kinds of purposes – successful or not. Now in so far as they realize all these purposes, the use of these devices does not necessarily result in calculating thinking, whatever other effects there may be in terms of distancing. Again, it is doubtful that there is a significant gap between the 'objective' world of calculation and 'subjective' experience in the case of contemporary devices. To the extent that they are all-purpose vehicles or nearly 'perfect' tools, it seems that the subjective is perfectly at home in them. If we manage to 'live' in them and through them, is there really always and necessarily a problem in terms of objectification? On the one hand, defending Simmel one could of course reply that this view is a sign that we have reached the final point of alienation and self-alienation. One could also, from a Marxian perspective, argue that our lifeworld is first turned into data, objectified and commodified, and then sold off to third parties (see also Fuchs 2011). In addition, one could argue that although we do all kinds of 'subjective' things with the devices, they still distance us from the immediate, embodied life experience and action. On the other hand, to this objection one could then reply that our experience is always mediated – that there is no unmediated experience since we always look through cultural glasses, and that online experience is as 'immediate' as offline experience. But how can we then explain experiences of distance and alienation?

My way of dealing with the latter problem (see also Coeckelbergh 2012a) is to make a difference between (a) a basic kind of 'existential' mediation by media and technologies and entanglement of subject and object, one which we cannot escape since intermediacy is part of the human condition (we are tool-making and tool-using beings, and there is no 'neutral', 'unmediated' view of reality) and (b) specific forms and degrees of alienation and distancing which are brought about, or made possible, but specific technologies, but which are at a different level of analysis than the basic, existential mediation that belongs to the human condition. In the latter kind of category (b) there is room, I think, to reflect on the 'extra' distancing created by financial technologies and, more generally, by electronic media and ICTs, which is what I do in this book. There may be no epistemic 'immediacy paradise' or social 'proximity utopia', there is always some distance because we experience through material and cultural glasses, but this does not mean that specific technologies such as money or computers cannot create an additional or rather *different* kind of distance. The physical, quantitative distance can be bridged. The question is whether there is a qualitative distance, in addition to and different from the basic existential distance that belongs to human being. Simmel is right to suggest that money creates such a qualitative distance (or a number of qualitative distances), and it seems plausible that electronic financial technologies and other electronic technologies do so as well – to different degrees. (Simmel's approach allows for degrees; see also below.)

But this account does not seem to solve the problem of the apparent entanglement of the 'objective' and the 'subjective' in, and made possible by, contemporary electronic technologies. To tackle that issue, I propose that we distinguish between different kinds of technologies, re-locate the (financial) technologies we are discussing in the context of specific practices, and then analyse how they make possible different degrees of alienation, objectification, and entanglements (or not) of the subjective and the objective. Then we could say, for instance, that highly complex financial technologies and the 'abstract' practices and 'calculative' forms of thinking and forms of life they are connected to, are indeed examples of objectification processes in modernity. They make possible and are part of processes in which there is little entanglement of the objective and the subjective, and they support processes in which the division of labour, the management, and the globalizing effects indeed produce more moral and social distance; processes in which the impersonal reigns. Many other electronic technologies and especially their specific *use*, by contrast, are much more part of the lifeworld – that is, they are part of other activities and contexts, including perhaps coin money – and are more 'personal'. Think about the use of mobile phones for personal communication and as "social" medium. It may well be that the hardware of these devices (and even part of the software) was originally created for other purposes, but now they are used differently and usually created for 'social' and 'personal' purposes, *rather* than for calculation. Although one can and must then in turn criticize this view – these devices are also used for objectification purposes by, for example, commercial companies and intelligence services, and they have distancing and disengagement influences – it is clear that Simmel made too little room in his analysis for the actual *use* and context of devices and made too little distinctions with respect to this use. The use of current electronic devices and its epistemic, social, and moral significance is far more ambiguous than I suggested in my interpretation and application of Simmel offered above. It is precisely their multi-purpose character as nearly 'perfect' tools, or at least as *more* perfect tools, that makes understanding and evaluating these devices so difficult.

Yet in spite of these objections concerning 'personal' devices and their use, with regard to financial technologies we can safely say that these devices – not the devices as such but their use in their context – contribute to making possible the objectification, commodification, and alienation of modern life in the context of globalization. While an analysis of, for instance, so-called 'social media' seems less straightforward, financial technologies in electronic form and in the context in which they operate today seem to be *distancing machines* indeed. Moreover, these effects add on to the basic existential distancing which is part of our being-in-mediation: there is always bridging and separating at the same time, and as tool-using beings who need tools to reach our aims and who can only become persons through the use of things, we are always living in an 'intermediate' and 'mediated' way: with, in, and by mediation.[8]

Concluding the discussion of this first problem, let me recommend that those of us who are inspired by Simmel's work on money (and I include myself here) must refine their conceptual tools to take into account the diversity of technological culture and practice and to allow for degrees in distancing and alienation. To use the language of McLuhan again: perhaps the medium has always been the message – this is my summary of Simmel's argument about the nature of money but it might be that in the context of modernity industrial and especially *electronic technologies* and electronic *financial* technologies become more 'perfect' media, more 'perfect' tools, creating more 'ideal' societies in Simmel's sense. It might be, in other words, that some media become *more* their message, become *more* pure tools, than others in specific contexts and under specific historical conditions. Today the medium of money, taking an electronic form, is more its message than any financial tool before. The corresponding history of distancing then goes as follows: there was always a basic existential distancing, but money already created new kinds of distances in the ancient world, under modern conditions the distancing increases, and contemporary financial technologies seem to be close to the final, highest stage in this development of distancing. Simmel's approach allows for degrees: we can say that electronic money has distancing effects 'in the same manner as other technical mediating elements, but does it more purely and completely' (p. 485). But we should not only consider variation in time (diachronical change) but also variation in practices and technologies at a specific time (synchronical change): there are also differences in the degree of distancing made possible by specific technologies and technological practices. This also needs to be analysed.

A second problem also concerns the broader cultural trend Simmel sketches, in particular the *dematerialization* (and hence spiritualization) thesis. Is it really true that the material becomes less important in the course of the history of our civilization? It is clear, for instance, that electronic forms of money and currency are less material and more 'spiritual' in this sense. This is also connected with the increasingly abstract forms of thinking in modernity. But at the same time these electronic tools, like all electronic ICTs, remain heavily dependent on material structures and infrastructures. For instance, in the financial world the *hardware*

---

8 Here mediation refers to both (1) medium as something 'intermediate' and (2) medium as 'environment'.

such as computers and servers used are *not* immaterial – even if the algorithms and the software is, to some extent – and the entire *infrastructure*, including for instance the glass fibre cables that connect financial centres across the globe, is very material indeed.

This example also connects to a third, related problem: the importance of *place*. My interpretation of Simmel's analysis suggests that modernity and global finance are so 'liquid' that place no longer matters, that there is only u-topia, non-place. But this is questionable in the sense that both the artefacts and the people involved in global finance are still located in specific places (e.g., financial centres) and most people still have personal and emotional ties with specific places, especially the places where they live and work. In this sense, the world of electronic, high-speed finance is far less 'ideal' or 'utopian' than suggested above. It seems that the relation between globalization and localization is far more complex. In Chapter 7, I say more about this, but consider for instance the example of high-frequency trading servers being placed in specific locations (if speed matters then place matters too) and the uneven geographical distribution of financial activities and power across the globe. Both the material dimension and geographical, 'local' dimension of global finance can and has been revealed by science and technology studies, geography, and related fields; within a purely Simmelian analysis they remain invisible.

Furthermore (this is a fourth problem), Simmel's analysis is unnecessarily *deterministic*. This point connects again to Simmel's history of objectification (with the growing gap between the objective and the subjective) and spiritualization. I do not wish to object to the very broad and general level of analysis Simmel uses. We also need other levels, for sure (see elsewhere in this section), but we also need hermeneutical building blocks such as the history of money and the history of spiritualization to better understand our modern culture and our civilization. I do not object to meta-narratives or 'Big Stories'[9], as long as we also have other stories and remain critical about the meta-narratives. What I wish to question instead, is the deterministic flavour of Simmel's (hi)stories. Consider again the story about the widening gap between the objective and the subjective, between technological development and cultural development. Simmel does not only argue that there is a gap between technological development and the development of the human mind (an interesting idea which is still very relevant today), he also suggests that the former is autonomous and takes its own course:

> If all the culture of things is, as we saw, nothing but a culture of people, so that we develop ourselves only by developing things, then what does that development, elaboration and intellectualization of objects mean, which seem to evolve out of these objects' own powers and norms without correspondingly developing the individual mind? (Simmel 1907, p. 449)

9 I refer to the discussion about postmodernism here.

On the one hand, I believe we should take seriously the experience and phenomenon this remark refers to. We indeed often have the feeling that the current technological developments, for instance in global finance, take their own course. It is tempting to say that when the developments of financial 'algorithms' continue, driven by the desire to (literally, perhaps) make money and profit, we are automating our way to a global financial-technological Singularity.[10] Perhaps there comes a point when financial artificial intelligence radically changes the world to such an extent that it becomes entirely unpredictable, out of control, inhuman or even non-human. Even if we reject this idea, it is important to examine the idea that 'technology takes over' (we must explore if there is some truth in it; automation has indeed problematic aspects), that technology always involves a tragic element, and that humans cannot fully control technological or social developments. On the other hand, as we can also learn from Simmel, money – and by extension all media and all technologies – are 'sociological', that is, 'human' phenomena, and if this is true then we are not entirely helpless or powerless since technology (as we know it) is not entirely 'external' to us. Technologies are part of our practices and societies, and practices and societies can be changed by humans, even if they are not completely controllable. To conclude, there is an 'autonomous' aspect to financial-technological development and even if we reject determinism and apocalyptic thinking about technology it is worth further examining this aspect, but since technological development is always also human-technological development we can at least try to steer the use and development of financial technologies in a more beneficial and ethical direction.

A similar answer can be given to Simmel's remark in 'The Crisis of Culture' that only the world of things is 'capable of unlimited refinement, acceleration and expansion, whereas the capacity of the individual is ineluctably one-sided and limited' (Simmel 1997, p. 92). I think that, on the one hand, it is both interesting and helpful to reflect on the limitations of our capacities in comparison to (at least some) capacities technologies. There seems to be a real problem here: because of the speed, abstraction, and complexity introduced by ICTs – including financial ones – it becomes harder to understand what is going on, let alone control it. On the other hand, it is doubtful that there really is such a (wide) gap between humans and technology. If tools and media are really social relations and social phenomena, as Simmel thinks, then this suggests that there is no autonomous, deterministic technological development, that we can change things – albeit not too much.

Fifth, Simmel's anthropology relies on the concept of *human essence*, which is problematic for various reasons; there is a long history of essentialism and anti-essentialism. For example, one could argue that natural evolution, societal context,

10 The Singularity hypothesis says that technological progress (mainly in AI, computing, and robotics) will result in triumph of non-human intelligence over human intelligence up to a point when human existence and civilization are changed radically beyond human comprehension, prediction and control.

and technological change influence what it means to be human and that it turns out to be impossible to provide an adequate definition of human essence without saying something about these influences. Both our ends and what it is to be human seem to be very dependent on our means and on the non-human. Therefore, there is no such thing as a human essence if this means a permanent, eternal Form. Furthermore, existentialists such as Sartre have argued that we do not have an essence, we exist: what we are and what the human is depends on what we make of ourselves. And elsewhere I have argued against what I called 'the properties view' in moral thinking, in particular in thinking about moral status (Coeckelbergh 2012b); this could also be read as an anti-essentialist argument. (However, I will not further develop these objections here.)

Sixth, so far my analysis is unconnected to media theory. In the next chapter on globalization and financial technologies I will use McLuhan and contemporary media philosophy. But allow me to already make one link at this point, since it adds a new insight to what Simmel says about money as a 'perfect tool' and what I have said about electronic ICTs as 'perfect tools'. A well-known and influential concept in (relatively) recent media theory has been 'remediation' (Bolter and Grusin 1999): new media refashion old media and "try" to become more transparent than previous media – if possible invisible. We have always tried to create communication technologies that are not themselves visible as medium, and each new medium has tried – in vain – to reach more transparency than the previous medium. It seems that this concept can also helpfully be applied to money and in particular to electronic money as understood within a Simmelian framework: perhaps electronic money tries to represent the exchange relation in a more transparent way than all previous, more substantial forms of money. By becoming information, data, and bits, money tries to reach immediacy; it wants to become invisible as a medium. It wants to build a bridge between people and between people and things, but it wants to be an invisible bridge. It wants to be better than gold bars and silver coins, which were all-too-visible as a medium because of their substance. It wants to be better than paper money, which is also still too material and which feels too much like a barrier in the exchange relation, too much like a material barrier between me and my purpose. Electronic money has the potential to become a bridge without resistance; a superconductivity express for exchange relations. But at the same time, through its abstract and symbolic nature, electronic money creates the greatest distance ever from the goods and people involved. Or to use the same metaphor: superconductivity requires the temperature to be lowered; this is the price to be paid. Moreover, the medium becomes highly visible itself when it becomes the ultimate end rather than the ultimate means. People no longer think about the goods they need or want; they want the numbers on the screen, they want the Bitcoins (the quantity of which is limited). The medium is fascinating, perhaps more fascinating than the message. (When I wrote the first version of this chapter, the value of bitcoins was rising, more than it was 'supposed' to do as a neutral, preferably transparent new financial medium. Now the situation is different, but it is likely that it will rise again or that there will be similar currencies in the near future.) Furthermore, electronic money as a 'perfect tool' cannot only take on all forms in terms of purpose; it is also a universal re-mediator: it can re-mediate all other financial technologies and 'comments' on them. For instance, Bitcoin can remediate coins and wallets in so far there are indeed (electronic) 'bitcoins' which people keep in (electronic) 'wallets', it can remediate cash in so far as it is anonymous and in so far as it can also exist in material form (coins with a code), and since it generally only appears on the screen it remediates electronic money on bank accounts. But to Bitcoin enthusiasts it 'feels better' than all these old media, since it is more fluid, more versatile, more the 'perfect' tool Simmel had in mind. It is part of a new monetary utopia. (See also Chapter 8.)

Finally, while Simmel has paid some attention to problems related to specialization of labour and while his view can be interpreted in terms of alienation, so far a more systematic political economy perspective is missing, also in my interpretation of his work. Although this is neither Simmel's nor my (main) approach, it is useful to interpret Simmel's view in the light of Marx's view of labour and his theory of money in order to further discuss the links between money, *labour*, and distance, and to reveal similarities between Simmel and Marx. A note on Marx may also be useful to highlight again the issue of *power*. This connects to the previous chapter but also to, for example, Chapter 8 in which I explore resistance and alternatives to existing social-financial practices and technologies.

For Marx money is a commodity which becomes a universal medium of exchange (Nelson 1999). What does 'commodity' mean here? Marx's focus is not so much on money as a mediator, a means of circulation, but on the relation between value, money, and labour. Money is itself produced by labour. It is a product or 'commodity'. Its value is determined by the cost of production, including the labour time necessary to produce it.[11] As what we may call 'the commodity of all commodities', money is also a symbol of commodity production for the market and therefore of alienation. Let us further explore his view and interpret it in terms of distance.

Like Simmel, Marx thinks that money has become an end in itself which absorbs all other ends. In 'On the Jewish Question' (1844b) Marx argues that money has become a worldly god, which degrades all other values:

> Money is the jealous god of Israel, in face of which no other god may exist. Money degrades all the gods of man – and thus turns them into commodities. Money is the universal self-established value of all things. It has, therefore, robbed the whole world – both the world of men and nature – of its specific value. Money is the estranged essence of man's work and man's existence, and this alien essence dominates him, and he worships it. (Marx 1844b, pp. 33–34)

11 We will see later that this gets an interesting twist in Bitcoin: bitcoins are produced by human and non-human labour.

Money is worshipped because, as Marx says in the *Economic and Philosophic Manuscripts of 1844*, with money we are 'capable of all that the human heart longs for, possess all human capacities' (Marx 1844a, p. 138). As a kind of value chameleon it can take on the value of all values. In so far as it is what Simmel called a 'pure tool', it promises to bridge every gap between ends and means. Marx writes that it converts our wishes from the realm of imagination to 'their sensuous, actual existence' (p. 139) and because of this mediation, this transformation from impotence to real power, Marx calls money 'the truly creative power' (p. 139). It bridges the distance between dream and reality, or at least promises that it has the capacity to do so. In Marx's view money thus has become more than a 'mere' instrument made by humans; as the essence of alienation it confronts us as something that stands at great distance from us. It is so distant that it becomes a god, an alien force that dominates us and that – in a truly monotheistic spirit – tolerates no other value.

This emphasis on money as an absorber of all value sounds similar to Simmel, but Marx's theory of money can and must be understood within his theory of alienation, especially alienation in the labour process. According to Marx, the main problem is not money, but alienated *labour*. As I have said earlier in this chapter, Marx argued that the division of labour renders the worker alienated from the product of his labour, from labour itself, from its species-being, and from others. Social relations become relations between things. What we may call 'money making' thus creates four types of distances, four types of alienation. It makes possible epistemic, social, and moral distances in the labour process and beyond. Money expresses these forms of distancing and is at the same time part of what makes them possible.

Money itself has also an alienated character in the sense that it is alienated from other commodities. There is an enormous distance with other commodities, and this makes it possible that it becomes an end in itself, that it is worshipped as the ultimate tool that makes dreams come true, that it is perceived as the ultimate *dream machine*. Again, according to Marx this is possible because money represents exchange as such (see also again Simmel). In *Capital* (Vol. I) he presents money as the expression of pure exchange value (Marx 1867). This makes it a commodity that is most alienated of all commodities: it is divorced from use value and alienated from labour. In Chapter Three, Section 2 of *Capital* (pp. 198–227) Marx writes that money is 'all other commodities divested of their shape, the product of their universal alienation' (p. 205). Money thus absorbs all other commodities and values, and it can do so because it is itself 'empty', it is pure function and in essence non-material. Like Simmel, Marx observes that in money function precedes over matter: 'its functional existence so to speak absorbs its material existence. Since it is a transiently objectified reflection of the prices of commodities, it serves only as a symbol of itself, and can therefore be replaced by another symbol' (Marx 1867, p. 226). Thus, as long as it has 'its own objective social validity' (p. 226), money functions and this function does not depend on materiality. It has 'a purely functional mode of existence in which it is externally separated from its metallic substance' (p. 227). This is in accordance with Simmel's insistence on the functional, social, and dematerial essence of money.

Thus, both Simmel and Marx pay attention to money becoming an end of its own, to money as pure function, and to alienation, but Marx analyses and puts more emphasis on the relation between money and modern *labour* and commodities in *capitalism*. For Simmel, by contrast, alienation and objectification are part of a broader process that is not restricted to what goes on in modern labour, production, and commodity exchange under capitalist conditions. It is a process that is not even restricted to *modernity*. Simmel sees distancing and alienation as phenomena inherent in the development of our civilization, which seems to move towards more objectification, intellectualization, rationalization, and calculativeness. Although Simmel knows that specialization of labour contributes to this process, he thinks alienation is not intrinsic to the workings of capitalism or to how commodities are produced and exchanged. Simmel shows that distancing comes in many forms; it is not only linked to labour and production. More generally, Marx focuses on the economic system, even if he sees implications for (all) values and makes comparisons to religion; Simmel's philosophy of money, by contrast, says much more about the wider social and cultural significance of money. Having said that, a Marxian perspective helpfully highlights the 'labour' aspect of money, and there are enough similarities between Marx and Simmel: both agree that money is essentially a non-material function, that money is about social relations, and that it tends to absorb other values and ends.

In addition, Marxian thinking in general can say more about money and *power* relations. I already mentioned power in the previous chapter, where I attempted a Foucauldian interpretation of the social and moral consequences of financial technologies, but clearly a Marxian perspective on power (e.g., one focused on class domination) can also be an important tool to better understand the relations between money and power, and money and politics. For instance, based on the *Economic and Philosophic Manuscripts of 1844* one could point to the power exercised by capitalists over labour, that is, over persons. Moreover, feminist approaches in this tradition have questioned 'masculine' abstraction and competition (Hartsock 1983). These topics are interesting in themselves and deserve much more attention. However, in this book I will focus on technology and distancing rather than technology and power.

To conclude, there are problems with Simmel's philosophy of money when it comes to its technology-human dualism (and indeed means-end dualism) and its suggestion that matter and place lose every importance in a spiritualized and globalized world. I have also shown that it has deterministic tendencies. It has also limitations with regard to accounting for the variety of (contemporary) financial media and technologies and the corresponding variety of use, context and practice. This also means it has problems to account for the variety of degrees of alienation and distancing. Moreover, Marxian thinking can bring a political economy perspective to the discussion and illuminate the relations between money, alienation and labour, which receives insufficient attention in Simmel. Finally, the discussion about Simmel's work may benefit from insights from other disciplines. In order to study 'technologies at work' – here: financial technologies at work – we need to connect Simmel's philosophy of money to work in the social sciences, for example concepts from media theory, empirical work about money in anthropology, studies of materiality and place in social studies of finance (including STS), and insights from financial geography. In the next chapters I will draw on some of this work.

Yet in spite of these problems and possibilities for improvement and development, Simmel's analysis remains an excellent starting point for thinking about the relation between financial technologies and distancing, in particular about the relation between on the one hand contemporary financial ICTs such as electronic money and currencies (but also screens, electronic trade platforms, electronic banking systems, etc.) and on the other hand problems of alienation, objectification, disengagement – problems of epistemic, moral and social distance. Like other critics of modernity Simmel suggests that these problems may well aggravate under modern conditions, but his main contribution lies in showing that these processes of distancing have their origin in technological-social developments that started much earlier in the history of humankind. It started with the financial technologies of ancient agricultural civilizations. It started also when we began to use tools that could not only better serve and mediate *specific* purposes, but that could also serve and mediate *more* purposes, a *variety* of purposes. In Simmel's analysis money is revealed as the tool of tools, the technology *par excellence*, and contemporary ICTs and media – financial ones and others – seem to fulfil this function of universal tool even better. They become better at converting our imagination into reality, they become even better dream machines. Moreover, Simmel also has shown that thinking about financial technologies must go hand in hand with thinking about our society – which is still a modern society. Thinking about financial technologies *is* thinking about society.

In the next chapters I will further elaborate my analysis, interpretation, and evaluation of contemporary financial technologies, and continue my 'dialogue' with Simmel. But after these histories and analysis of distancing processes in history, we now need to start shifting our attention to the present and future of financial technologies and their relation to distance. We need a more developed analysis of new forms of money and, more generally, *contemporary* financial technologies. For this purpose I will continue my discussion as an inquiry in the field of *philosophy of technology*, although I will also draw in other disciplines. I will also connect Simmel's ideas to twentieth century philosophy of media and technology (in particular McLuhan) and to problems in practical philosophy (in particular moral responsibility). Moreover, I will consider objections to Simmel from contemporary anthropology (e.g., Hart 2007) and explore the literature in social studies of finance. This will also help to deal with the problem that Simmel's analysis remains at a rather general level and it is important to also consider the phenomenology of money and financial technologies at the level of specific practices and contexts. But I will continue the 'dialogue' with Simmel: in Chapter 4, I will study the relation between financial ICTs and globalization (here I will make a connection between Simmel and McLuhan); in Chapter 5, I will further develop the 'metaphysics' and 'sociology' of electronic money (again I will reconnect also to Simmel here); in Chapter 6, I will further discuss the moral implications of contemporary financial technologies by analysing the relation between financial automation and moral and social responsibility (I already mentioned quants and high-frequency trading in this chapter); in Chapter 7, I will 'localize', 'materialize', 'humanize' and 'personalize' finance and consider objections to Simmel from anthropology; and in Chapter 8, I will explore alternative financial technologies and – keeping in mind Simmel's lessons concerning the entanglement of money and the social – alternative forms of social order.

# Chapter 4
# Geography 1: Financial ICTs and the global space of flows

## 4.1. Introduction

In the previous chapters I discussed various forms of distancing and alienation made possible by money and other financial technologies. It seems that today this development has reached a (provisional?) peak: money has become entirely insubstantial and mediates impersonal trade relations that span the globe. Money and other financial technologies seem to contribute to growth and maintenance of a global sphere of electronic virtuality, to what Castells called 'the space of flows' in his work on 'the information society' (Castells 1996; see also below).

Simmel has shown that this process already started when money became detached from materiality; now the dematerialization of finance – and therefore of society – seems complete. I say 'seems', because in a later chapter I will use current social studies of finance (anthropology, geography) to question this geography, and attend more to the local dimension of global finance and to the relation between financial technologies and place. It will turn out that place is not obsolescent; even in 'global' electronic times place is still very important to human experience and action – also in finance. Moreover, in the next chapter I will question the virtual/real and digital/non-digital distinction. But let me now first further construct the general distancing and alienation story as a conceptual tool to make sense of broader financial-social developments in the history of our culture (and indeed civilization). In this chapter I will pay particular attention to the geographical dimension of these developments and their distancing effects.

For this purpose I present a discussion at the intersection of geography and philosophy of media and ICTs (in the next chapter I will add metaphysics). First, I look into the connection between technology and globalization. I say more about the definition and theory of globalization (Robinson 2007; Bartelson 2000; Scholte 2002), and highlight the social-spatial dimension of globalization. I argue that in order to study this dimension of globalization and its relation to financial technologies, we need to better understand technologies and media. I draw on McLuhan's *Understanding Media* for this purpose. I also further discuss the relations between space and technology by referring to Arendt's discussion of what she calls 'world alienation' and 'earth alienation' (Arendt 1958). I make connections with Simmel.

Then I focus on more contemporary technologies, in particular electronic ICTs. I rely on the work of Castells which connects globalization to the rise of the 'network society' and the role of ICTs in this process. I explain how there have been space-time compressions as a result of ICTs (see for example Brey 1998). I highlight the role of information technologies in shaping the 'global space of flows'. I discuss the role of screens and networks in shaping global and 'virtual' financial markets (Cetina & Bruegger 2002a and 2002b) and argue that our culture has changed as a result of globalization (e.g., Allen and Pryke 1999 – they also engage with Simmel).

In order to explore the moral and social implications of these phenomena, I also rely on (other) philosophy of technology and media in the context of globalization, in particular McLuhan's approach to media and his view of the relation between globalization and responsibility, and of course his interpretation of *money* in *Understanding Media* (McLuhan 1964). Again I make comparisons with Simmel (and Arendt). Connecting Simmel to McLuhan, I support again the main 'distancing' thesis I construct and critically discuss in this book: *although financial technologies, like all contemporary ICTs, are very good in bridging physical distance, they at the same time create a number of epistemic, social, and moral distances*. This seems especially true in the age of electronic technologies. It turns out that both Simmel and McLuhan sketch a history of increasing detachment and abstraction. For instance, we live in a world of numbers which alienates us from objects and from others. It becomes more *difficult to see how responsibility is possible under such conditions*. However, I also show that McLuhan interprets contemporary electronic technologies in a different, perhaps more optimistic way: in this age, when the world becomes one and we see the consequences of our actions, it is far more difficult to escape responsibility. Moreover, McLuhan suggests that the new technologies create at least the possibility of a more integral and involved existence, one that is in some ways similar to pre-modern existence. If this is true, maybe financial technologies can play an important role in revealing interconnectedness and relationality to us, and in rendering us *more* rather than less involved.

## 4.2. The global space of flows

### *4.2.1. Globalization as a spatial and social process: Changes in social space*

If we want to understand the epistemic, social and moral aspects of contemporary financial electronic technologies, we need to connect our analysis to discussions about globalization and its relation to ICTs. Let us start with the concept of 'globalization'.

Although there is no single accepted definition, most of us will agree that – to use Robinson's helpful definition – globalization refers to 'increasing connectivity among peoples and countries worldwide' *and* to *awareness* of these interconnections, an awareness of the world as a single place; the concept thus has an 'objective' and a 'subjective' dimension (Robinson 2007, p. 127;

see also Robertson). These connections and the awareness of them go beyond 'inter-national' connections or connections between people from different continents; they change physical, social, and political reality. It is important to see globalization as a transformative process rather than an interaction between pre-existing units (Bartelson 2000). Globalization changes people, social institutions, states. It even changes their conditions of existence, as Bartelson argues, since it '*despatializes* and *detemporalizes* human practices as well as the conditions of human knowledge' (Bartelson 2000, p. 189). We do things differently, but we also experience the world in a different way. The epistemology changes, our knowledge and experience changes: we take a global viewpoint. We come to see things differently.

Furthermore, globalization can refer to a process or to a condition, or to both: it is usually assumed to be a process which leads up to the present condition which is more 'global' than ever. That present condition is seen as fundamentally different from the past. The idea is that of course there have been 'global' connections before. For instance, the *internationalization* of money and finance is a process that began much earlier in history (Lothian 2002). In Chapter 2 I have also suggested that 'globalization' already started in the ancient world. And arguably there has been a 'global' imagination in, for instance, Greek philosophy. However, recently transplanetary links are denser than those of any previous period in history (Scholte 2002, p. 16) and our awareness of a 'global' world has increased. It is also worth noting that, as described in the globalization literature, histories and theories of globalization often connect the concept of globalization to other key concepts such as capitalism, modernity, internationalization, westernization, colonization, or indeed the network society (see below). Globalization is linked to other processes in society and culture.

The concept of globalization has an important spatial dimension. The idea is that there is a change in the relation between space and social structure, that there is *a change in social space*. Scholte proposes a conception of globalization which explicitly identifies globalization in those terms. Globalization is about 'the spread of transplanetary … connections between people' (Scholte 2002, p. 13) and this change constitutes 'a shift in the nature of social space' (p. 14). Globalization is about spatiality *and its relation to human experience and to the social*:

> The term globality resonates of spatiality. It says something about the arena of human action and experience. In particular, globality identifies the planet – the earthly world as a whole – as a site of social relations in its own right. Talk of the global indicates that people may live together not only in local, provincial, national and regional realms, as well as built environments, but also in transplanetary spaces where the world is a single place. (Scholte 2002, p. 14)

The spatial aspect of the social life should not be taken as a marginal fact about it; instead, as Scholte rightly says it is 'a defining feature of social life'; any description of the social is incomplete with a 'spatial component' (p. 14). This means we should talk about 'spatial-social' change rather than treating geography and sociality as separate spheres. For the discussion about the meaning of globalization, it means that we must ask the more precise question: what, exactly, is the nature of the *spatial-social change* indicated by the term 'globalization'?

One could argue that it means that there is now 'one world', that we experience our world as 'one world'. But this, by itself, could also mean 'one national world', 'one local world', etc. Scholte argues that there is a 'shrinking world', but this shrinking does not happen *within* a territorial space (e.g., the territory of the nation state) but beyond it (Scholte 2002, p. 19). There is time-space compression, but because it is global, it abolishes the territory and calls into question territory, distance, and remoteness itself (Scholte 2002, p. 19). As Heidegger puts it in his essay 'The Thing':

> All distances in time and space are shrinking. … He now receives instant information, by radio, of events which he formerly learned about only years later, if at all. … The peak of this abolition of every possibility of remoteness is reached by television …. Yet the frantic abolition of all distances brings no nearness; for nearness does not consist in shortness of distance. … Short distance is not in itself nearness. Nor is great distance remoteness. … What is happening here when, as a result of the abolition of great distances, everything is equally far and equally near? What is this uniformity in which everything is neither far nor near – is, as it were, without distance? Everything gets lumped together into uniform distancelessness. (Heidegger 1971, p. 163–164)

This is a very specific meaning of 'one world': a world where there is no distance. Heidegger also distinguishes between physical distance and the experience of distance as remoteness or nearness. What counts here is the phenomenology of distance; the point is not that there is no longer physical distance, but that this physical distance and other distances are experienced differently – indeed that 'distance' as such is experienced differently. Moreover, it is interesting that Heidegger connects this shrinking and indeed abolition of distance with technological developments: in this time this was radio and television. We will need to say more about technologies and media below. Let us first further discuss the concept of globalization and what it does to (our experience of) space and time. We can find further ideas about what happens to time-space under modern conditions in the 'classic' literature on globalization (and on modernity). Giddens, for instance, has argued that the essence of globalization is 'time-space distanciation': under (late) modern conditions, the level of time-space distanciation is much greater than in previous (e.g., agrarian) civilizations (Giddens 1990, p. 14). Today we witness 'the intensification of worldwide social relations which link distant localities in such a way that local happenings are shaped by events occurring many miles away and vice versa' (Giddens 1990, p. 64). In this book on postmodernity Harvey has argued that capitalist development has produced globalization as representing 'time-space compression' which reduces the constraints of space (Harvey 1990).

And Sassen has argued that globalization is based on a network of global cities (Sassen 1991). However, her work suggests that the 'global' is still linked with specific places. I will say more about this in Chapter 7.

For now it is important to note that in all these definitions and theories of globalization the *spatial dimension* is central, that what matters is our *experience* of space and distance, and that there are significant links between spatial and *social* structure and dynamics, between the physical and the social, between technological and social development, between on the one hand the physical and the material, and on the other hand the social and cultural. The distanciation and the compression have to do with human experience and have social consequences or, better, are at the same time 'physical'/'material' and 'social'. Geographical and human, social transformation are entangled. There is not 'social' change next to 'spatial' change; there is spatial-social (or social-spatial) change and social-geographical change. This is also important to keep in mind in the next sections when I discuss the role of *technologies* in relation to globalization (and finance), in particular financial ICTs. Using Arendt, McLuhan, and also more recent authors such as Castells, I will discuss the relations between space and technologies in general, but I will pay special attention to ICTs and in particular financial ICTs.

*4.2.2. ICT, globalization, and finance (McLuhan 1)*

Technology has always played a major role in the history of globalization and in the distancing processes that are related to it. But as Heidegger's remark on radio and television already suggested, modern information and communication technologies – including financial ones – deserve a special place in the growth of global interconnections and the awareness of being part of one world. Some globalization scholars explicitly mention the role of ICTs. For instance, Scholte recognizes that if 'contemporary world history is supraterritorial to degrees well beyond anything previously known' (Scholte 2002, p. 21), this is also due to ICTs such as the telegraph (nineteenth century) and to satellites, TV, and internet (twentieth to twenty-first century). But given the intended but especially the *unintended* influence of technologies, the philosophical significance of this link between ICTs and globalization can hardly be overestimated, and deserves further discussion before we focus on the role of *financial* ICTs.

On the one hand, it is obvious that electronic technologies and digital media contribute to globalization. They are meant to bridge distances, and they have made possible webs of communication that span the globe and interconnect people. They bring the distant (distant people, distant things, distant worlds) closer to us. Or at least this is their aim; this is what they are designed for. At the same time, however, this physical bridging has psychological, epistemic, social, and moral effects and meanings that go beyond what they were intended for. They change not only how we communicate and with whom we communicate; the medium also becomes the message, to use a famous phrase from McLuhan. Through the medium, by using the technology, we act in different ways and we perceive the world in different ways. McLuhan suggests that in the end these non-intended effects of media turn out to be more important for the history and development of our culture and civilization that any 'content'. Let me further explain his view.

In *Understanding Media* (1964) McLuhan explains his famous expression: to say that the medium is the message means that 'the personal and social consequences of any medium – that is, of any extension of ourselves – result from the new scale that is introduced into our affairs by each extension of ourselves, or by any new technology' (McLuhan 1964, p. 7). I will not, at this point, further discuss McLuhan's definition of technology in terms of 'extension'; what is important for now is (1) to recognize that there is an important *spatial* dimension to this technological change and its personal and social consequences (a change of scale) and (2) to recognize and understand, like Simmel and McLuhan, that technologies are always *social* technologies: they alter 'our relations to one another' (p. 7). How do the new electronic ICTs alter the social?

In the previous chapter, I presented Simmel's view that money leads to social distancing. Interestingly, however, *McLuhan* thinks that automation technology, in contrast to earlier 'machine' technology, is 'integral and decentralist in depth' rather than 'fragmentary, centralist, and superficial in its patterning of human relationships' (p. 8). McLuhan's view of the consequences of technology for human relationships thus differs from Simmel and Marx, and is based on his interpretation of new *electronic* media rather than industrial machines. (I will return to this in my discussion of McLuhan's interpretation of money.) Yet in terms of approach he is entirely in line with Simmel, Marx, and (other) philosophers of technology when he argues that the medium is the message. It is worth saying more about this approach, since it is so crucial for understanding and constructing the argument of this book.

McLuhan is not interested in the purpose of media and technologies, what they are used for, their 'content'. Instead, he wants to study 'the psychic and social consequences' of technologies, the – largely unintended – effects of them. McLuhan criticizes the (still) common view that science or technology is not good or bad in itself, but that there are only good and bad uses (p. 11). Against this view, he says we must not ignore the nature of the *medium* and suggests that we should realize how much technology is connected to what we are as humans. In McLuhan's idiom: we need to study not so much the content but the 'grammar' of technologies (p. 14). Technologies and media go hand in hand with different kinds of 'perception and organization of experience' (p. 17). As individuals we are not in (full) control of these changes. Someone may say 'I pay no attention to ads' (today we may say internet, mobile phones, Twitter messages, etc.), McLuhan argues, but the effects are of a different kind, of a kind perhaps only the most sensitive among us – such as artists – can fully experience. Whatever one may *think* about technologies like the internet, with McLuhan we must understand that they have effects at a different level:

> The effects of technology do not occur at the level of opinions or concepts, but alter sense ratios or patterns of perception steadily and without any resistance. The serious artist is the only person able to encounter technology with impunity, just because he is an expert aware of the changes in sense perception. (McLuhan 1964, p. 19)

However, not only perception alters; a technology or medium has also social and political effects, which are directly related to patterns of perception. At this point in this text McLuhan gives the example of 'the operation of the money medium' (p. 20): he describes how money in seventeenth century Japan caused the breakdown of feudal government, opened up the country, and 'reorganized the sense life of peoples' (p. 20). Importantly, he says that this change in the lives of people 'does not depend upon approval or disapproval of those living in the society' (p. 20). According to McLuhan, the impact of technology on society is not something that can be decided by agreement; the medium itself brings about changes to our experience, existence, social relationships, and political order. This is true for media in the sense of ICTs, but also for all kinds of technologies. McLuhan writes:

> Anybody will concede that society whose economy is dependent upon one or two major staples like cotton, or grain, or lumber, or fish, or cattle is going to have some obvious social patterns of organization as a result. ... Cotton and oil, like radio and TV, become 'fixed charges' on the entire psychic life of the community. And this pervasive fact creates the unique cultural flavour of any society. (McLuhan 1964, pp. 22–23)

Thus, the effects of technologies and media have to do with the medium itself. This also means: the effects have to do with the media they – to use a contemporary term – re-mediate. McLuhan already says that 'the "content" of any medium is always another medium' and claims that electric light is 'pure information' (McLuhan 1964, p. 8). (I will return to the issue of information below.)

Furthermore, these effects can and must also be described in *spatial* terms (in the text cited previously McLuhan talks about a 'new scale' that is introduced by technology), and like the phenomenon of globalization the effects have both objective and subjective aspects. Once again the assumption is that culture/mind is strongly related to physical space and to matter.

For instance, when today we consider ourselves as part of one world, one globe, and one planet, this is made possible by material technologies and media such as satellites and images on computer screens through which we literally *see* one globe, one planet earth. We perceive the world as global and act 'global' not only because we have understood abstract arguments and have received information; our global perception, thinking, and action is made possible and actively shaped by contemporary ICTs which, in this case, create a distance between us and the world, enabling us to see the world as a globe and as 'one'. (McLuhan and others would use the word compression; I will return to this.) More: our ICTs do not only enable and facilitate this perception as mere instruments; they also promote it and shape it. The means also shapes the end. As a medium, ICTs are tools that transmit a particular 'content' or 'message', but as McLuhan argued they are also the message, and as such they have social-spatial effects.

Distancing (and bridging) between people and between humans and objects is one such effect; it is related to money (as Simmel argued) and to other technologies (e.g., transportation). It is also connected with the effect of compression and implosion (McLuhan), which is created by the new technologies and media. But before I say more about compression and implosion, let me first discuss – in the line of my reading of Simmel – one particular process of distancing which is relevant to the discussion about globalization and which is also made possible by technologies and media: distancing from the earth.

### *4.2.3. Distance from the earth: Hannah Arendt on alienation*

In the previous section I wrote that contemporary ICTs create distance between us and the world, and I suggested that this distancing and alienation enables us to see the world as 'one', as a 'globe'. Technological change and change in thinking thus went hand in hand. However, as Hannah Arendt's work suggests, this particular process of distancing – distancing from the earth – already started earlier: at the time of the birth of modern thinking and modern technology.

In *The Human Condition* (1958), Arendt tries to understand the modern age and writes about what she calls 'world alienation' and about how science and technology made this possible. She writes that 'the invention of the telescope and the development of a new science that considers the nature of the earth from the viewpoint of the universe' (Arendt 1958, p. 248) set in motion what we may call with McLuhan *a different pattern of perception*: this technology and science has resulted in 'the shrinkage of the globe' until 'each man is as much an inhabitant of the earth as he is an inhabitant of his country'. (Arendt 1958, p. 250) The consequence of these technological and scientific changes is that distance no longer matters. According to Arendt, 'speed has conquered space' and distance becomes meaningless:

> Men now live in an earth-wide continuous whole where even the notion of distance, still inherent in the most perfectly unbroken contiguity of parts, has yielded before the onslaught of speed. Speed has conquered space; and though this conquering process finds its limit at the unconquerable boundary of the simultaneous presence of one body at two different places, it has made distance meaningless, for no significant part of a human life – years, months, or even weeks – is any longer necessary to reach any point on the earth. (Arendt 1958, p. 250)

Consider again what Heidegger said about the abolition of distance: 'everything gets lumped together into uniform distancelessness'. But if there is no distance, then we are everywhere and nowhere. Our body may be still physically located at one place, as Arendt suggests, but our presence is no longer tied to a specific location. The meaning of the very words 'location' and 'presence' are put into question when our 'presence' is 'global', spans the earth. (Note that today such simultaneous presence is possible through the mediation of ICTs. I will return to that soon.) But what makes Arendt's analysis really interesting is that she shows that the emergence of this 'globalized', 'one earth' condition and perception has a history, a *material*, technological history, which is at least partly not within human control. Of course the explorers of the early modern age did not *intend* to abolish distance; they wanted to bridge it. Yet Arendt says that this is precisely what they did with their technical inventions: their maps, instruments, and (later) technical inventions made it possible that 'all earthly space has become small and close at hand' (p. 250). Thus, Arendt thinks there was already a shrinkage of space before those technical inventions, made possible by the 'new science' which considered everything from the viewpoint of the universe, but also by the process of increasing abstraction in our thinking – scientific and otherwise. And as Simmel also observed, this shift towards more abstract thinking has a lot to do with *numbers* and other 'tools' of the human mind such as models. They enable us to bridge the distance between the 'small', localized and embodied human understanding and the 'big' globe and the vast universe which initially was beyond the reach of our mind:

> Prior to the shrinkage of space and the abolition of distance through railroads, steamships, and airplanes, there is the infinitely greater and more effective shrinkage which comes about through the surveying capacity of the human mind, whose use of numbers, symbols, and models can condense and scale earthly physical distance down to the size of the human body's natural sense and understanding. Before we knew how to circle the earth, how to circumscribe the sphere of human habitation in days and hours, we had brought the globe into our living rooms to be touched by our hands and swirled before our eyes. (Arendt 1958, p. 251)

However, bridging this distance comes at a price. If we are concerned with the remote, we may lose our epistemic, social, and moral entanglement with, and engagement with, what is near us. Arendt argues that the 'shrinkage' of the earth meant that we were alienated from our immediate surroundings, removed from the earth. The bridging of one distance thus resulted in a new kind of distance; we become less bound to the earth, we become less earthly:

> It is in the nature of the human surveying capacity that it can function only if man disentangles himself from all involvement in and concern with the close at hand and withdraws himself to a distance from everything near him. The greater the distance between himself and his surroundings, world or earth, the more he will be able to survey and to measure and the less will worldly, earth-bound space be left to him. The fact that the decisive shrinkage of the earth was

> the consequence of the invention of the airplane, that is, of leaving the surface of the earth altogether, is like a symbol for the general phenomenon that any decrease of terrestrial distance can be won only at the price of putting a decisive distance between man and earth, of alienating man from his immediate earthly surroundings. (p. 251)

Again we must note the role technology played and plays here. The invention of the aeroplane (and later space technology) did not only enable us to physically leave the earth; it also resulted in changes to our perception and thinking. Arendt's analysis suggests that today our thinking 'flies'; when we developed flight technologies, our thinking and imagination also developed the capacity to fly. Alienation from the earth is thus the result of new technologies and corresponding changes in the human imagination. Another technology Arendt mentions is the telescope and the science related to it. She writes about 'the great boldness of Copernicus' imagination, which lifted him from the earth and enabled him to look down upon her as though he actually were an inhabitant of the sun' (p. 259). Again this meant that we became alienated from the earth, and indeed from nature. With today's science and technology we 'handle nature from a point in the universe outside the earth', and 'have found a way to act on the earth and within terrestrial nature as though we dispose of it from outside, from the Archimedean point' (p. 262).[1] By formulating universal laws, we have become universal beings: 'creatures who are terrestrial not by nature and essence but only by the condition of being alive, and who therefore by virtue of reasoning can overcome this condition not in mere speculation but in actual fact' (p. 263).

Natural science thus had (and has) an important role in this process of abstraction and what we may call epistemic alienation. Mathematics freed humans from 'the shackles of spatiality', from the shackles of earth-bound experience' (p. 265). With algebra 'terrestrial sense data and movements' were reduced to 'mathematical symbols' (p. 565). These symbols enabled man to place nature 'under conditions won from a universal, astrophysical viewpoint, a cosmic standpoint outside nature itself' (p. 265). Arendt calls this, with a spatial term, the 'condition of remoteness' (p. 267).

Arendt's analysis can be applied to contemporary ICTs, which seem to boost the alienation processes identified. Contemporary ICTs literally enable us to look at the earth from a distance, but they also seem to create conditions under which the world further shrinks and distance is abolished, conquered by speed. Electronic devices and equipment, created by a science that is as abstract and symbolical as one can possibly imagine (information science), enable us to take distance from our immediate surroundings, connect us instantly with the most 'remote' places, and make possible simultaneous presence. The result is that we

1 The Archimedean point is a hypothetical vantage point, by which one removes oneself from the object of inquiry; Archimedes wanted to lift the earth standing on a point outside the earth.

are alienated from our direct environment and that our culture becomes alienated from the earth: thinking, perceiving, and acting as Arendt's modern scientists, we look at everything from the vantage point of the universe and are disentangled from what is or what is near us. Place, it seems, no longer matters. It seems as if the 'local' is abolished.

In our 'electronic age' or 'information age', financial technologies seem to contribute to this process: ICTs in trade and electronic monies make possible globalization as interconnectedness, but they also create distance. ICTs at the same time bridge and abolish territory and distance; they have both alienating and universalizing effects. They enable global flows of money and seem to create one social and epistemic space, one global space. How can we understand this? Let me first say more about ICTs in general, and then further discuss financial ICTs in particular.

### *4.2.4. The information age and the network society: the space of flows, shrinking and blending*

In his famous work on the information age and 'the rise of the network society', Castells argued that we live in a new age of information due to the development of new information technologies, which create a new economy that is 'informational', 'global' and networked (Castells 1996). It is informational because the productivity and competitiveness of economic agents 'fundamentally depend upon their capacity to generate, process, and apply efficiently knowledge-based information' and it is global because its core activities and components are 'organized on a global scale' (Castells 1996, p. 66). It is an economy that involves global financial markets and the globalisation of trade and technology. It is networked because productivity and competition play out in global business networks. Castells argues that all this is made possible because of 'the information technology revolution', which 'provides the indispensable, material basis for such a new economy' (p. 66). We have information economies. Moreover, it is crucial to understand that information does not only contribute to this process but becomes *itself* the product of the production process; the new technology transforms 'all domains of human activity' (p. 67). We now produce information.

But there are also spatial (and temporal) features and consequences of the new technologies and the new society Castells describes. In Chapter 6 of his book, Castells argues that both space and time are transformed (Castells 1996, p. 376). He makes a distinction between the 'space of flows' and 'the space of places': the first concept represents the 'new spatial logic' connected to the new technological system, whereas the latter refers to 'the historically rooted spatial organization of our common experience' (pp. 377–378). Let us for now focus on this 'new spatial logic'. Let us accept that there are global information flows, global networks and processes. Cities become informational cities, some become megacities. But what exactly is the 'space of flows'?

According to Castells, there is a meaningful but complex relationship between society and space. Like Scholte and other theorists of globalization, Castells thinks space is social and the social is spatial. Space 'is society' since spatial forms and processes are formed by social structure and since social processes act on the inherited built environment; therefore, space 'cannot be defined without reference to social practices' and it is created by material practices and processes (Castells 1996, pp. 410–411). What Castells calls the 'space of flows', then, is a new spatial form which emerges in the network society and consists of electronic impulses (based on information technology; this is the material basis), nodes and hubs (the network links up to places), and people – in particular the dominant, managerial elites – the social actors (pp. 412–515). Thus, it is a network of technologies and social actors, of things and people.[2] Now although Castells stresses that most people live in places, that is, perceive their space as place, he thinks that the logic of the space of flows is dominant and sees a gap between these two forms of space: there are two 'parallel universes whose times cannot meet because they are warped into different dimensions of a social hyperspace' (p. 428).

Are these indeed 'parallel universes'? Castells' view must be understood in the context of the first decade of the internet, when the internet still appeared as constituting a separate world (cyberspace), a 'virtual' world separate from the 'real' world. It the 1990s it seemed that there was a gap between the 'online' and the 'offline' world, which corresponds to the gap between on the one hand a (then) new symbolic environment (the internet), a virtuality, a 'space of flows' and on the other hand a material and local environment, a 'space of places'. This gap also corresponds to what Castells calls 'the Net' and 'the Self'. It seems that in the early twentieth-first century we have moved to a different view, perhaps because today information technologies tend to encompass all other forms of media – see also what I said about Simmel and remediation – and because we have become 'Inforgs' (Floridi 2007). To put it differently: to the extent that the 'space of flows' has conquered the 'space of places', or to the extent that both 'universes' have merged, the gap has become less relevant. Thus, Castells' assumption about the parallel universes must be questioned. At the same time, what Castells calls the 'space of flows' does refer to an important aspect of the global and technologically mediated experience we have today. Whatever our local attachments may be, as users of electronic media we may have the experience that we are 'in' a space of flows, that we are part of its flows. If there is 'global' tendency in the social-spatial processes we witness, live, and experience, then 'shrinking', 'interconnectedness', 'distance', but also 'flows' seem to be suitable descriptions of our new experiences – descriptions that are in line with the work of Simmel, Arendt, and other critics of modernity. Even if there is only one informational 'universe' and if we are 'Inforgs', then the experience of being part of that informational universe may be described as being part of information *flows*.

2 With Latour one could call this a network of actors and 'actants'. However, in this chapter I will not discuss Actor Network Theory or related work.

More generally, it is undeniable that the new information technologies have changed both society and the spatial structures that are relevant to it, and the current form of globalization cannot be understood without taking into account the information and communication technologies that have emerged in the past decades. Castells and others have theorized this. However, it is also clear from this brief exploration of the relations between globalization, technology, and space that there is a need for further analysis of the spatial processes connected to globalization and electronic technologies. Before moving on to global finance and financial ICTs, let me therefore further discuss the 'shrinking' argument.

Some of the previous arguments can be connected to what Brey, influenced by Giddens and others, has called 'the geographical disembedding thesis': places are less and less determined and defined by physical-geographical features, which has altered the identity of places and even the very concept of place (Brey 1998). Places are spaces that are meaningful and are related to the social life and to identity. But because of the new ICTs, it seems that geographical features become less important and geographical distance matters less. As already suggested before, there has been 'shrinking' or 'time-space' compression. Distance has been abolished (see again Heidegger and the next section on McLuhan) and global interdependence has increased. Now in order to call attention to the *abolishment* of distance rather than shrinking, Brey also uses the term 'blending': if electronic media abolish time-space barriers then there is no longer a distance, then it is indeed possible that one is in two places at the same time, since in one's experience both places have *blended*. This brings us once again to McLuhan, who not only suggests an interesting approach to technology but who also offers an original view of *globalization* and its relation to new technologies and media, human experience, social relations, and even responsibility.

### *4.2.5. Experience and responsibility in the global village (McLuhan 2)*

In *Understanding Media*, published decades before the rise of the internet, McLuhan wrote that 'the Western world is imploding', that 'we have extended our nervous system itself in a global embrace, abolishing both space and time as far as our planet is concerned' (McLuhan 1964, p. 3). McLuhan argued that using new media, we extended our senses and nerves, and now we're living in what he famously called a 'global village': we live in one large, global technological and psychic complex. According to McLuhan, this globalization of consciousness also involves an attitude of detachment, which started in earlier ages, but he argues that in what he calls the 'electric age' and in the age of what we today call 'globalization' such an attitude is more difficult to maintain:

> In the electric age, when our central nervous system is technologically extended to involve us in the whole of mankind and to incorporate the whole of mankind in us, we necessarily participate, in depth, in the consequences of our every action. (McLuhan 1964, pp. 4–5)

Thus, according to McLuhan, the new media, by interconnecting the experience, actions, and consequences of these actions of the whole of humankind, have a different moral effect than the one observed by Simmel. Simmel argued that the development of more connections to those far away from us happened at the expense of indifference to our immediate surroundings. McLuhan (and Castells and many others), by contrast, argues that everything, 'local' and 'global', the near and the remote, becomes one. McLuhan does not think this new epistemic and psychological condition leads to indifference, but quite the opposite. To the extent that our world really globalizes, indifference is no longer possible. What happens at the other end of the line is, if we follow McLuhan, as much part of our body and our senses as what is near to us. The result is that we also feel involved and responsible. Thus, according to McLuhan in a globalized world there is not *more* 'moral distance' but *less*.

However, there are a number of problems with this view. I see the following barriers to the kind of moral awareness McLuhan envisages and to the globally responsible action that might follow from it (see also Coeckelbergh 2013a). First, global responsibility may be hard to bear, and it is questionable whether, even if we are involved in 'everything', we also know everything. There are psychological and epistemic limits: even if we wanted to, we cannot carry the world upon our shoulders. Although the new technologies have undoubtedly increased our knowledge and participation, we are neither omniscient nor omnipotent, and we can only take up *partial* responsibility for what happens in the world. Second, limited knowledge also applies to what we know about our precise contribution to what happens in the world. Epistemic work is needed, an active effort on our part to know what goes on, to know our contribution to it. But does the internet show us 'the world' and 'everything', or does it show us only a part of it, perhaps even only 'my' world? This danger is especially present in contemporary social media and in search engines, which show us a world adapted to our preferences and to the communities we live in. The 'other' might remain out of sight. Third, there are limits to what we can do. Even if we were in touch with everything that happens in the world and knew how it related to our actions, there are limits to individual agency. It is unclear how much we can do as individuals, and it is unclear if for the world McLuhan describes a different model of action can be used. Furthermore, McLuhan neglects the issue of power: some people know more than others and can *do* more than others. He suggests that the position of social groups may change for the better (see below), but the social-critical part of his analysis remains underdeveloped. Finally, instead of action the new technologies may only stimulate our imagination but fail to make possible real engagement with the world. It may stimulate an 'aesthetic', detached attitude. For instance, inspired by Kierkegaard, Dreyfus has argued that the internet allows play and exploration but lacks ethical engagement and commitment (Dreyfus 2001). Whether or not this is true, it is an important issue to discuss and McLuhan does not really address it. More generally, the techno-mystic dimension of his vision (which I think is inspiring) sometimes seems to take precedence over his invitation to study and reflect on the more concrete 'personal, political, economic, aesthetic, psychological, moral, ethical, and social consequences' of new media, to use McLuhan's own words in *The Medium is the Massage* (McLuhan & Fiore 1967, p. 26).

Thus, in these senses there is still 'moral distance' due to our limited capacities when it comes to the psychological cost of bearing 'global' moral responsibility and problems in terms of knowledge and agency that it raises. Nevertheless, McLuhan shows that in terms of moral awareness and responsibility there is also a positive side to the new world we live in, and that the new technologies and media may open up new possibilities – including new psychological, epistemic, social, and *moral* possibilities. As Heidegger said: 'where the danger is, grows the saving power also' (Heidegger 1977, p. 34). Let us further explore these new possibilities and this 'saving power', but now with a focus on the main theme of this book: *financial* technologies and distancing.

For financial ICTs McLuhan's view means that in principle it becomes possible, through the technologies, to feel responsibility towards those far away, who bear the consequences of one's financial-technological actions. Perhaps electronic money, for instance, could act as a 'moral communication technology', a medium that does not breed indifference but exactly the opposite. By further interconnecting us, it could reveal the relational character of the world, and thereby remind us of our moral responsibility. Our actions and our bodies have been globalized; now we are desiring and awaiting the globalization of moral awareness and moral responsibility. However, this interpretation is not only problematic in principle for the reasons given above; it is also highly questionable if this effect takes place to a sufficient degree under current conditions, that is, given the form and practice of current financial technologies. Responsibility has at least two conditions of possibility: we need to be in control and we need to know what we are doing. But it is questionable whether current financial ICTs and other ICTs support these conditions of responsibility. For instance, there seems to be an epistemic (and therefore also moral) gap between on the one hand the world of the trader, that is, his knowledge and experience, and on the other hand the people affected by his trade decisions. If a person in London or New York trades agricultural goods, for instance, does (s)he really know what the effects are for people at the other end of the world? Moreover, it is questionable what 'his' or 'her' trade decision means when increasingly *machines* take over. I will say more about these problems of responsibility in Chapter 6 on trade automation technologies (e.g., high-frequency trading).

Yet according to McLuhan responsibility follows once there is 'compression': he thinks that once we have electronic technologies that give us an enhanced awareness of our global interconnectedness, we also have an enhanced awareness of our global responsibility due to the compression by electronic technologies. This is his argument:

> After three thousand years of specialist explosion and of increasing specialism and alienation in the technological extensions of our bodies, our world has

> become compressional by dramatic reversal. As electrically contracted, the globe is no more than a village. Electric speed in bringing all social and political functions together in a sudden implosion has heightened human awareness of responsibility to an intense degree. It is this implosive factor that alters the position of the Negro, the teen-ager, and some other groups. They can no longer be contained, in the political sense of limited association. They are now involved in our lives, as we in theirs, thanks to the electric media. (McLuhan 1964, p. 5)

This is a far more optimistic message than Simmel's: as the globe implodes the social and political life, existence under electronic technologies is not the summit of distancing and alienation but its end. McLuhan connects 'electric technology' to 'the aspiration of our time for wholeness, empathy and depth of awareness' (p. 5). He writes about 'a faith that concerns the ultimate harmony of all being' (p. 6). Let me therefore ask again: could financial technologies help to bring this new awareness and even harmony about, rather than alienating us from our surroundings and pushing us into the desert of indifference? In order to tackle this question, let us first explore again in what sense(s) finance is globalized (this section), and then further discuss its potential epistemic, moral, and social effects (which I will do at the end of this section and especially in the next section).

Finance has certainly been globalized in the senses discussed in the previous sections. Financial technologies helped to shrink the world and to blend spaces, or at least they helped to shrink *the world of finance*: there seems to be only one global financial space left. To talk about finance today is to talk about global finance. Furthermore, financial technologies such as electronic money are very much part of a more general abolishing of distance and place, and of the culture of speed that Simmel, McLuhan, and also for instance Virilio (1977) considered to be characteristic for modernity. Electronic money also seems to dematerialize culture: as I have said in my application of Simmel to contemporary ICTs, money in electronic forms becomes a mere symbol or idea. The culture of speed is a culture of dematerialization, which, paradoxically, is brought about by technologies. Technology is always considered to be 'material'. But technologies and media themselves have dematerialized. In our globalized world, there are new rhythms and modes of interaction, and contemporary finance seems to contribute to that new technological culture. Thus, there is no doubt that finance has been globalized and globalization has been financial.

This claim is also supported by more empirically oriented literature. Cetina and Bruegger, for instance, write about the global 'lifeform' of financial markets. Screens and networks shape global and 'virtual' financial markets (Knorr Cetina & Bruegger 2002a and 2002b), which take on a life of their own. For instance, according to Knorr Cetina, global currency markets are 'by all accounts … genuinely global markets. As collective disembodied systems generated entirely in a symbolic space, these markets can in fact be seen as an icon of contemporary global high-technology culture' (Knorr Cetina 2005b, p. 38). She argues that global currency markets and similar financial markets are 'flow markets': there is 'a "melt" of material that is continually in flux' (p. 40). She shows that these trading floors and markets are 'heavily dependent on electronic information and communication technologies' (p. 44). Some markets are even *entirely* electronic. Moreover, for the functioning of all these markets, computer screens and terminals are very important since they act as teletechnologies that 'deliver to participants a global world' (p. 41), projecting financial reality, which today is always a global reality.

The 'electronic' character of these markets and of global finance does not mean that they are entirely immaterial. As I will note in Chapter 7, they still need a material infrastructure, including material things such as networks, cables, satellite connections, trading floors, terminals, and screens. But it seems that in terms of how the financial world appears to us – including as we will see to its participants or inhabitants – this material infrastructure is only secondary; it seems that global finance is more than anything else a world of flows. It is not so much a world of objects; in the 'electronic' age and the 'information' age it is first and foremost a world of processes and *information*:

> What discloses itself to participants in the mass of materials on their financial screens is not the presence of objects but the presence of information. What we are really dealing in, traders say, is information. ... What shows up on their screens are not 'beings' at all but rather moments of opportunities to act that pass quickly .... Thus traders find themselves in a succession of shared informational situations or 'clearings'. The mundane economic meaning of an informational reality that opens itself is that it discloses opportunities for investment and speculation. (Knorr Cetina 2005b, p. 42)

Thus, to the traders themselves their world does appear as an information world. What is disclosed, then, is not 'truth' but 'news' (p. 43). Furthermore, Knorr Cetina argues that what was meant as a medium becomes a being on its own:

> The mirrored market that is comprehensively projected on computer screens acquires a presence and profile on its own, with its own properties. Traders are not simply confronted with a medium of communication through which bilateral transactions are conducted, the sort of thing the telephone stands for. They are confronted with a market that has become a 'life form' in its own right, a 'greater being', as one of our respondents ... put it. (Knorr Cetina 2005b, p. 47)

Thus, in terms of knowledge and experience traders live in a world that has become *distant* from the humans that make it possible. It is an anonymous world (p. 47), which then appears to the traders as an alien being. Consider again what Marx said about alienation (see Chapter 3). In the *Economic and Philosophic Manuscripts* of 1844, he wrote that 'the worker is related to the product of his labour as to an alien object' (Marx 1844a, p. 71). And in 'On the Jewish Question' he said about money that it is 'the estranged essence of man's work and man's existence' and that 'this alien essence dominates him, and he worships it' (Marx 1844b, pp. 33–34). Thus, although 'the market' is the product of the traders' labour and although its flows of money are related to *human* work and *human* realities, it appears as a 'greater being', a god which is worshipped.

Moreover, if the market and, more generally, global finance appears to those who participate in it as an alien being, this also means that it is 'out of our control'. It has its own presence and, we might say, its own *agency*. (I will not further discuss this thought here, but it is worth further unpacking.)

Note also that it is not only participants in the markets who experience 'the market' as an alien being, a god to be worshipped. Numerous people around the world feel that they depend on 'the market' for their well-being and indeed survival, and therefore *also* experience 'the market' as an alien force and/or a divine being. Especially in times of crisis (when most of this book was written and conceived) most citizens feel that they are in the hands of an all-powerful being called 'the market', 'global finance', or 'the economy'.

The idea of a 'space of flows' seems also highly applicable to the world of global finance and its market-spaces. If the market is a 'greater being', it is also one that continuously transforms itself, it has 'ontological liquidity' and is therefore perceived as what Knorr Cetina calls 'reality in flux' in contrast to the durability and materiality of the physical world (p. 52). She also refers to Castells' idea of a network society based on flows of information, but stresses that what is transferred is not actual money but rather financial power, which is (even) more abstract (p. 53). Furthermore, there is also what Giddens, Brey, and others called *disembedding*: 'the markets observed appear removed from their local context in terms of participants' orientation, their inherent connectivity and integration as the key to overcoming the geographical separation between participants', and other features (p. 56). Thus, although globality does not mean the total absence of place(s) – as I will argue in Chapter 7, reality is more complicated and this form of globality depends for instance on 'bridgehead centres of institutional trading in the financial hubs of the three major time zones' (Knorr Cetina 2005b, p. 57) – the metaphors of 'flow' and 'flux' seem to be very suitable to express how participants and observers experience global markets and their geography, indeed our experience of global finance in general.

Now what are the implications of this financial globalization for (the rest of our) culture? It is clear that our culture and society have changed as a result of (financial) globalization, and that these changes should be understood in social-spatial terms. But what, exactly, has changed? Is it mainly the (experience of) alienation we can discern here? Or can we interpret some developments in terms of more 'moral awareness' and 'harmony'? How did and does global finance change patterns of perception and experience and patterns of social life? In the previous chapter on Simmel and earlier in this section I already started discussing and interpreting what happens to culture and society in a world globalized by electronic technologies. Now that I have said more about financial globalization we need to continue this journey. One way to do this, and also to revisit the very question we are trying to answer, is to return to *Simmel and McLuhan*. Let me briefly touch again upon Simmel in this section, and then move on to the next section where I will continue my reading of McLuhan.

Although Simmel is surprisingly absent from much contemporary literature on financial globalization, a few authors explicitly refer to his work and use it in order to better understand contemporary finance and its effects on our culture. For instance, in their article on money cultures Allen and Pryke use Simmel as a starting point to say something about the relation between on the one hand the mobility of money and things and on the other hand meaning and identity (Allen and Pryke 1999). They argue that today's money markets influence our ways of experiencing space and time. For instance, they claim that derivatives 'pull distant spaces into centres of rhythmic coordination which coordinate exchange in a new form of monetised space-time' (Allen and Pryke 1999, p. 52) and that this world of circulation shapes the 'spatial logic' of daily life, that it influences how people 'give meaning to their surroundings and to others' (p. 54). What, then, is this influence? The authors first mention three kinds of effects described by Simmel: (1) the ability of money to overcome distance and the resulting distancing in social relationships, (2) changes in the rhythm and fluctuations of social life, and (3) a shift in the pace and acceleration of modern life (p. 55). They then apply this Simmelian framework to contemporary financial instruments. For instance, in the money culture of the modern city there is social distancing, and Allen and Pryke suggest that new forms of money such as derivatives also have this effect. However, the authors then take distance from what they take to be Simmel's view and stress that there is no single experience (it depends on context) and that there is always a process of imagining and meaning-giving. With the latter they mean that it is not the space-shrinking technologies and digital flows *themselves* which make our lives faster and which make us more mobile, but rather – and this is my interpretation of their view – that the acceleration and mobilization of our life and culture is made possible by the technologies *and* the meaning-giving, the technologies *and* our imagination. Thus, human subjectivity should be part of the picture. The Simmelian argument I constructed so far should not be read as saying that the technologies 'cause' specific effects in society and culture, as if we are part of a deterministic machine with no room for other experiential (and action) possibilities. Human experience is an active process of meaning-giving and imagining. And, as Allen and Pryke rightly suggest, this means that different impressions, interpretations, and meanings are possible. (I will return to this aspect of meaning-giving in Chapter 7. See also alternative visions of finance and society in Chapter 8.)

Thus, contemporary financial technologies raise once again the question concerning the psychological, epistemic, social, and moral implications of globalization. This is especially true for financial ICTs, for new *electronic* financial media which create streams of electronic money and electronic information. What kind of (social) spaces are created by these new technologies? Do we really live in a 'global village'? What is the role of electronic money in these social-spatial developments? Has money become information? Is that the end stage of the Simmelian process of dematerialization? And does it mean that *we* become information? In order to further discuss money as medium, its relation to (other) ICTs and to information, and its influence on society and culture, let us take a closer look at McLuhan's view of money and compare it with Simmel. In the next chapter I will say more about the 'metaphysics' of money.

### *4.3. Money as medium and the movement of information (McLuhan 3)*

McLuhan's interpretation of money as a medium enables him to compare it with other media, an exercise which is very interesting. Earlier in this chapter I reviewed McLuhan's view that media change our sense of life and society – regardless of our agreement or disagreement – and his view on the relation between globalization and technology. So what does that mean in the case of money, and financial technologies more generally? How did and do they change our sense of life and society in the context of globalization and the information age?

On the one hand, McLuhan's analysis approaches Simmel's and Arendt's view of the relation between distancing and modernity, and their view that there is an increase in abstractness. In particular, McLuhan seems to share Simmel's view that distancing is a process that started long before the modern era. In Chapter 14 of *Understanding Media*, McLuhan compares the development of money with the development of speech:

> Speech comes with the development of the power to let go of objects. It gives the power of detachment from the environment that is also the power of great mobility in knowledge of the environment. So it is with the growth of the idea of money as currency rather than commodity. Currency is a way of letting go of the immediate staples and commodities that at first serve as money, in order to extend trading to the whole social complex. (McLuhan 1964, p. 143)

This idea corresponds with the thesis that money – like in fact other teletechnologies – at the same time bridge and distance. As Simmel argued, if and when money becomes currency, it creates distance between people and things. Trading is extended, and so is the distance between humans and objects. We 'let go' of them. Thus, put in spatial terms there is indeed bridging and distancing. In line with his view that technology is about extending the human body and senses, McLuhan argues that money extends our grasp from nearest commodities to distant ones, and compares this with 'extension' in the development of children and greater apes:

> Just as the hand among the branches of the trees learned a pattern of grasping that was quite removed from the moving of food to mouth, the trader and the financier have developed enthralling abstract activities that are extensions of the avid climbing and mobility of the greater apes. (McLuhan 1964, p. 144)

Furthermore, McLuhan stresses that money is intrinsically linked to the life of the community, to the social life. Like Simmel, he defines money as 'a social medium or extension of an inner wish' (p. 147). And like Simmel, he interprets the change from commodity money to representative money as a process of increasing abstraction. He writes that when money became paper money, it lost its aura: 'Just as speech lost its magic with writing, and further with printing, when printed money supplanted gold the compelling aura of it disappeared.' (McLuhan 1964, pp. 145–146).

Like Simmel, McLuhan thinks that these changes in the form of money influence the form of our social relationships and work, and more generally what we may call our form of life. In describing the function(s) of money, he focuses on the spatial effects, in particular on bridging and distancing. It is a bridge and at the same time action at a distance: ' "Money talks" ' because money is a metaphor, a transfer, and a bridge. ... It is action at a distance, both in space and in time' (p. 147). Thus money is something relational and a medium of communication like language. As such it influences how we think, live, and work. Its development is linked to the history of increasing specialization of labour (see also Simmel and Marx) and to what we may call the compartmentalization of modern work and life. McLuhan writes that money

> separates work from the other social functions. Even today money is a language for translating the work of the farmer into the work of the barber, doctor, engineer, or plumber. As a vast social metaphor, bridge, or translator, money – like writing – speeds up exchange and tightens the bonds of interdependence in any community. It gives great spatial extension and control to political organizations, just as writing does, or the calendar. (p. 147)

McLuhan thus connects the social and epistemic influence of money to the social and epistemic influence of literacy, which also made possible more abstract thinking and specialization of labour (p. 148). Both go hand in hand, or even stronger: the written word made possible the world of money. 'The written word' includes numbers, which make possible markets and prices: 'the world of prices and numbering is supported by the pervasive visual culture of literacy. Nonliterate societies are quite lacking in the psychic resources to create and sustain the enormous structures of statistical information that we call markets and prices' (p. 148). McLuhan thus interprets money in the context of the history other technologies and media, and discusses the changes in perception, thinking, and culture that are connected to that history. Considered from this perspective, money is not so 'special' as we might think (and neither are its effects very special). Like Simmel, McLuhan shows that the development of money is part of a larger cultural-technological development – not only in modernity but also in the history of civilization – towards more abstraction and distancing. It is 'just' another distancing technology, a medium which like many others contributes to various forms of distancing. For instance, our pricing system replaces earlier forms of exchange, which required much more personal involvement (to use a more Simmelian expression). McLuhan gives the example of haggling: 'The extreme abstraction and detachment represented by our pricing system is quite unthinkable and unusable amidst populations for whom the exciting drama of price haggling occurs with every transaction.' (McLuhan 1964, p. 149). Like Simmel, McLuhan thinks that money influences social relations as well as 'the inner life': it changes economic transactions and social relations, but it also changes our way of *thinking*. Let us start with economic transactions and social relations.

McLuhan discusses money in the context of what we would now call globalization, or more precisely the globalization of human relations: new ICTs, new media have made possible 'the instant electronic interdependence of all men on this planet' (p. 149). We are all connected by global information flows. Electronic money plays a key role in this. For instance, it changes work and trade. The result of the global information flows that have grown during the past decades is that work becomes 'the sheer movement of information' (p. 149). The form of money changed from a material form to pure information: 'From coin to paper currency, and from currency to credit card there is a steady progression toward commercial exchange as the movement of information itself.' (McLuhan 1964, p. 149).

However, when it comes to the effects on division of labour, McLuhan departs from Simmel. He thinks that today the division of labour changes, but in a rather different way than Simmel described. Instead of a further increase and intensification of the division of labour, something else happens. He argues that in tribal societies there was no work in a strict sense, there was no specialization of labour. In agricultural communities (and, I would add, in industrial societies) this changes: there is division of labour. McLuhan: 'Where the whole man is involved there is no work. Work begins with the division of labor and the specialization of functions and tasks in sedentary, agricultural communities.' (p. 149). With Simmel we could say that in industrial times this division is further increased and refined. McLuhan too thinks that there is a process of increasing fragmentation. Interestingly, however, he thinks that with electronic technologies this changes again. The fragmentation stops: 'In the computer age we are once more totally involved in our roles. In the electric age the "job of work" yields to dedication and commitment, as in the tribe.' (p. 149).

Is McLuhan right about this? On the one hand, electronic media are used in the service of labour, paid labour we do as employees. In this context, new media seem entirely compatible with division of labour and (other) features of the persistent modern organization of labour. It is not clear if division of labour is weakening. For instance, contemporary organizations use ICTs and new media, but they seem to be used to maintain rather than diminish specialization: at the level of the person, we do not find 'the whole man' but only part of what we (think) we are or want to be. We remain employees. On the other hand, it seems true that electronic media demand 'total involvement' and maybe the line between 'work' and 'leisure' time is gradually disappearing. As users of social media, for instance, we do unpaid work as producers of data that companies can sell off to third parties.

But we do not experience this as 'work'. We may call it work-leisure or leisure-work. And the new technologies enable us to work at home, if needed 24hours a day, where we mix work and leisure. The same medium (for instance internet via a smartphone) is used for both kinds of activities. Perhaps in this sense today our involvement is indeed more 'total', more like the work of the members of a tribe. Rather than modern workers, who specialize, we become hunters and gatherers again. McLuhan indeed thinks we engage in 'information gathering' (p. 150).

A significant difference that McLuhan does not discuss, however, is that the earlier type of hunters and gatherers were not alienated from the product of their work and worked for themselves, not for a company (paid or unpaid). Even if we were totally involved when we use electronic media, as information processors we have no control over the information we sell or 'leak'. We have to be good in many things rather than being specialized, but to the extent that we work for a salary or that we are slaves in the data mine of large companies all these skills and all that knowledge are used for a 'product' that is alienated from us. Moreover, if we are always involved in work-leisure, it seems as if we never have a 'break'. Having a 'break' becomes meaningless. Is this a problem, or not? There are many issues to discuss here. Nevertheless, McLuhan seems right in suggesting that what I called our 'work-leisure' appears to us as hunting and gathering rather than, say, planting, breeding, or manufacturing. The world of information and communication looks more like a wilderness than a farm or a factory. We are totally involved in new ecologies of information.

This has further implications for how we experience our lives and for how we *think*. As we hunt and gather for information, we do neither 'work' nor 'play'; we live in an informational world and moving around in that world becomes increasingly more obvious and trivial, and is no longer questioned – just as the members of a tribe do not question their world. For us to talk about 'online' or 'offline' may become as meaningless as it was for pre-modern people to talk about 'work' or 'leisure', and perhaps this is already the case today. Such dualistic categories have no place in a more integral understanding of human existence, an understanding which itself seems to be propelled by the new media and technologies. Our technologies do not only follow thinking (as 'applications' of our thoughts, concepts, laws, symbols, etc.); thinking also follows our technologies.

Financial technologies may also play a role in this development towards a less modern and more integral form of life. McLuhan highlights the role of money and ICTs in the change in work and existence: first money fosters specialization, but then the technologies become more 'integral' and 'organic'. He writes: 'Nowadays, with computers and electric programming, the means of storing and moving information become less and less visual and mechanical, while increasingly integral and organic.' (McLuhan 1964, p. 150). For financial technologies, this means that the flows of information and money make possible a world that is 'offline' and 'online' at the same time: we are never really 'outside' of global finance (or 'inside' for that matter). We are always part of streams of money-information, we help to produce financial data, and we hunt and gather financial information. Our financial life and other lives only *partly* play out in the bureaucracies and mechanical worlds sketched in the first chapters; they also play out and develop in an environment that feels more like the integral world of the hunters and the gatherers. This means that if McLuhan is right we do not only experience distancing; in so far as new financial and other electronic ICTs shape our world, there is also place for involvement and commitment, for 'wholeness', and perhaps for tribal community.

To conclude, there are at least two things we can learn from McLuhan here:

First, there is something valuable in McLuhan's method. In his analysis McLuhan treats money like any other medium; he thinks that this has to be done since 'any study of one medium helps us to understand all the others. Money is no exception' (McLuhan 1964, p. 151). The story of money is also the story of other technologies and vice versa. I already suggested this in the beginning of this book when writing a brief history of financial technologies: the story of money is entangled with the story of other financial technologies and indeed other technologies *tout court*.

Second, in terms of his 'message' there are striking similarities to Simmel. Although with McLuhan we can call attention to new, electronic forms of money and distinguish these from earlier-modern forms in terms of its social influence and significance (e.g., on work), much of what he says about money bears striking similarities to Simmel's story about dematerialization and distancing. Like Simmel, McLuhan first sketches a story of how money becomes increasingly what Simmel argued it always has been: an agent of exchange. McLuhan continues and updates this story for the electronic age, the age of speed and the abolishment of distance, and there is a significant difference with Simmel with regard to his history of the division of labour, but the general story line remains roughly the same: 'As money separates itself from the commodity form and becomes a specialist agent of exchange (or translator of values), it moves with greater speed and in ever greater volume.' (McLuhan 1964, p. 152). Money is thus seen as part of a technological, economic, and cultural history in which technologies and media such as writing and the alphabet shaped society and its values. It is also part of the story of globalization as propelled by technology: 'paper money enabled Western industry to blanket the globe' and translates work 'from one culture to another' (p. 153). Today it seems that electronic money makes possible economic globalization and the 'translation' of work. (In Chapter 7, I will nuance this claim: labour is still rather territorial.)

Also particularly interesting for the analysis of electronic financial technologies today, is McLuhan's view (already mentioned before) that money is not only a transmitter of information, but that it *is* information. One might also say: it becomes *communication.* This means it becomes what Simmel called a 'pure tool', it becomes exchange itself. This has implications for other goods and values. As McLuhan (also) says, money reduces everything to a common denominator (p. 155). The latter sounds similar to Marx, but the difference is now that 'information' is now the key concept. In contrast to Simmel and Marx, McLuhan gives us more conceptual tools to reflect on the role of contemporary (financial) *information and communication* technologies. This includes issues concerning automation (a topic which will be discussed in Chapter 6): towards the end of his chapter on money McLuhan makes an explicit point about automation technology: ‘Money, which had been for many centuries the principal transmitter and exchanger of information, is now having its function increasingly transferred to science and automation’ (p. 154). McLuhan thus thinks that (material) money is being *replaced* by electronic technologies; this claim is interesting in view of the next two chapters, which include discussions about Bitcoin and the issue of automation. In any case money and electronic automation technologies become *connected* in the electronic age, especially in the form electronic money and electronic currency. Contrary to many other commentators on money, McLuhan understands that there is a tight connection between money and technology, money and media. He says that today everything comes to have an ‘informational aspect’ (p. 154). Contemporary electronic money is thus part of what some call the information revolution (see also again Castells). If money can and must be understood as a medium and a technology, then today this means that contemporary forms of money must be understood as contemporary media and technologies, that is, as contemporary *information and communication* technologies. McLuhan teaches us that like other technologies and media these are always active *as* media: they transform our senses and our perception, our social relations, and our culture and civilization.

What is then the ‘message’ of contemporary electronic ICTs, including financial ones? Both Simmel and McLuhan define our culture as a culture of numbers and a culture of speed. Today money must be understood as part of the streams of information that blanket the globe, creating a world of abstraction and distance, but also, perhaps, a world that has the potential to become more organic and integral. It might become almost as organic and integral as the natural world – if such a distinction will make sense at all to the next generations.

But if it is true, as McLuhan argues, that today we are not less but *more* involved (at least more involved than in the agricultural and industrial age), what kind of involvement is this? How can and will we (as humans, individuals, and societies) cope with the new global electronic culture that is made possible by new forms of money and by other electronic technologies and media? How are we already coping with it? If we are going to find answers to these questions, it is important to learn from the writings of Simmel, Arendt, and McLuhan that thinking about globalization and finance means also thinking about technology and that re-thinking and evaluating new forms of money means also, and perhaps mainly, re-thinking media, technologies and society. In the end, it means re-thinking human experience and human existence; it means re-thinking our world.

In the next chapters I will further discuss both Simmel and McLuhan when relevant. I will also further develop objections and discussions already briefly mentioned in this chapter. For instance, what exactly is the relation between money and automation? What does automation in finance imply for responsibility? And is there still 'a place for place' and a place for materiality in the world of global finance? Is everything 'remote' in the financial global village?

In the next chapter I will reflect on the meaning and 'metaphysics' of money, and connect the phenomenology of money sketched so far to more 'general' phenomenological epistemology: what we say money 'is' depends on our world, that is, on our perception and interpretation of reality, including our perception of 'what is'. I will focus on the 'ontology' of money and the role of subjectivity and the social in what money 'is'. Questions in this domain already figured in the chapter on Simmel but also in this chapter when the question concerning the nature of money was touched upon and when dualistic modern assumptions were questioned. What is money? What does it mean to say that money is an object? What does it mean to say that money is social? What is the social? What does it mean to say that money is information? Are information technologies and the world they create indeed becoming organic? Are they no longer artificial then? Is there a separate kind of 'cyberspace' (and hence a separate kind of financial cyberspace), as many authors argued or assumed in the 1990s, or do we have to radically rethink reality in the 'information' age, a reality which is always 'our', human, reality? But does living in the information age mean that *everything* is information? Is it dangerous to reduce everything to 'information'? Many questions can be asked here and I will not have the space to discuss all of them (in depth). But I think it is important for a good discussion of financial technologies in the electronic and information age that we at least ask these questions and that we realize that the question regarding financial technologies is related to fundamental philosophical questions about reality, about things, about relations, about technology, and about the human.

# Chapter 5
# Bitcoin and the metaphysics of money

## 5.1. Introduction

In the previous chapter I argued that money and other financial technologies have created a global space of flows, a phenomenon that seems to amount to the dematerialization of finance. It appears that money and trade have become increasingly detached from the physical, material world, that this development has created various kinds of distancing (including 'moral distance'), and that ICTs have played, play, and most likely will continue to play a key role in this distancing. However, in Chapter 7 I will nuance this picture, because to understand distancing in the electronic age it is important to get a clearer idea of the nature of financial ICTs and their ethical and social implications. For this purpose, it is particularly important to understand and reflect on the meaning of money today, that is, the meaning of money in the 'electronic' age or the 'information' age. What is money in the electronic age? What does it mean? Is it a thing? Is it something 'digital'? Is there only one meaning or are there more? What kind of 'flows' are these global financial flows? Are these streams of money, of information, of numbers, of what? Indeed, given that numbers play such an important role in finance, we also need to say something about their meaning and status. What are numbers, and what is the relation between numbers and money? And what are the ethical and social implications of money in all these meanings?

The question concerning the meaning of money and numbers viewed in the light of processes of globalization and dematerialization becomes especially urgent in the case of electronic money. Contemporary computer-based and network-based information and communication technologies have made possible new forms of money, including new forms of currency. This impacts the lives of human beings. Partly directly, for example when we perform financial transactions on the computer and we pay electronically, but also indirectly, since today the finance sector is entirely dependent on electronic ICTs and what happens in that new, transformed financial world has a significant impact on society. In order to evaluate that impact, we need to better understand new forms of money and the new ICTs. What is it the finance sector trades? What is this thing called 'money', which seems to be 'over there' at exchanges or 'up there' in global streams, and which influences our existence? Is money a 'thing' at all? Is electronic money a 'thing', and, if not, what is it? If we want to know what money and numbers are doing to us, we want to know more about their nature and meaning. We need a 'metaphysics' of money and in particular electronic money.

Moreover, the question concerning the metaphysics of money seems especially relevant today as we see the emergence of electronic currencies that have no equivalent in the 'material' world; they only seem to exist 'on the computer', 'in the network', 'in the streams of information', etc. They are perhaps 'digital' or 'electronic' forms of money in the strictest or purest sense, since it seems that they have no counterpart or basis in the physical, material 'offline' world. For example, Bitcoin is a peer-to-peer payment network which involves no central authority. It is usually regarded as a digital currency which only 'exists' in 'the digital world'. Bitcoins are generated by a computer algorithm. It has no 'real', material-physical coins or banknotes. First it seemed only a game for computer nerds and for 'early adopters'. However, after having been introduced 6 years ago it is now much more widely used. It has attracted speculators and people who engage in illegal activities, for example for black market purchases or money laundering (due to the perceived anonymity), but increasingly Bitcoin is also used for payment of legal products and services and for speculation. Interestingly, like many internet services it is a non-state controlled, decentralized system. (I will return to this issue.) Consider also currencies used in computer games (in-game currencies) and virtual worlds, which equally lack any material equivalent or physical form. For instance, *World of Warcraft* works with Gold and *Second Life* has Linden Dollars. They are merely 'virtual' currencies that are part of 'virtual economies within games, although there are links to the 'real' economy (there are exchanges between 'virtual' currency and 'real' currency). Note that here too state control and regulation is limited.

Yet in spite of their 'digital', 'electronic' and 'virtual' nature, these forms of money raise ethical problems, for example about security and privacy (this was already the case with earlier forms of electronic money and financial ICTs, for instance electronic bank transfer) or use by criminals or authoritarian regimes. Thus, it seems that their nature is supposed to be 'digital' or 'virtual', but their consequences are 'real'. For instance, we consider some 'virtual' activities to constitute a 'real' crime (e.g., theft of money deposited on an electronic bank account), or we exchange 'virtual' money into 'real' goods or 'real' currency. But if this is true, then are these electronic forms of money also 'real' themselves? What is 'real'? Again we cannot avoid metaphysics questions (which I will approach from a phenomenological perspective).

This chapter offers a conceptual framework that organizes and guides reflection on the meaning of money, numbers and other financial technologies that seem to belong to the 'electronic', 'digital' world, to the 'virtual' financial streams and currents that span the globe. First, it offers a preliminary ontology – later interpreted as a phenomenology – of money, electronic money, and other relevant entities in global finance and their relation to technology. For instance, I discuss the nature of Bitcoin. I also propose what I call a 'deep-relational' view of money. Then I further discuss the relation between money and technology, and between money and (social) institutions. I draw conclusions for (thinking about) the ethics and politics of electronic money.

## 5.2. The metaphysics of 'electronic' money

Questioning the nature and meaning of financial technologies is not very common because usually we do not think about them. We are so used to money, for instance, that normally we do not reflect on it. We use it. But do we know what we are doing and what we are dealing with? In the *Confessions* Augustine asked: 'What then is time? If no one asks me, I know what it is. If I wish to explain it to him who asks, I do not know.' The same can be said about money, or numbers. When we use financial technologies, most of the time they appear as what Heidegger calls 'ready-to-hand' (German: *zuhanden*) (Heidegger 1927): although we sometimes talk about them in a way that presupposes concepts and theory (e.g., concepts from financial and economic theory), we generally use them without theorizing and without thinking. Money is seldom 'present-at-hand' (German: *vorhanden*), it seldom appears to us as a thing or concept to think about. Only when there is a problem we may think about it. Philosophers and other thinkers, however, reflect on things in the 'present-to-hand' mode. For example, they ask: 'What is money?' and (as I also do in this book) '*Where* is money?' In this section I continue my inquiry into the metaphysics and geography of money, with a particular focus on electronic and 'virtual' money.

In the first chapters we have encountered a number of meanings money can have, for example money as a medium of exchange, a medium of communication and information, a social relation, an institution, and of course a *technology* – an information and communication technology. If one needs *one* definition at all, I propose the latter definition is able to capture a lot of the other meanings, since technologies function as media, are social in their nature and implications, and many of them contribute to information and communication. This is why in this book I write about money as a 'financial technology'. However, in this section I want to emphasize the variety of meanings money can have and explore how they are connected to more general (or other) ways of experiencing the world. To use the language of metaphysics: I propose to distinguish between a number of *ontologies* that all have implications for what money 'is'. In the language of phenomenology this means: a number of ways money and other financial entities can appear to us. I prefer the latter approach and I will say more about it later, but I will start with a straightforward question concerning the nature of money: what is money?

To answer this 'is' question and to answer similar 'is' questions regarding electronic money and Bitcoin, we need the branch of metaphysics that deals with such 'is' questions: ontology. I propose that we distinguish between the following ontologies of money which each have implications for the 'being' or 'nature' of electronic money and electronic currency (e.g., Bitcoin).

### *5.2.1. Object ontology: Money as a thing, electronic money as a 'digital' or 'virtual' thing*

Many people experience 'what is' in terms of objects, things. 'The world' then appears as a collection of objects, different kinds of objects. We categorize objects. For instance, 'things' can refer to 'natural' things or 'artificial', human-made things (artefacts). We also distinguish between 'material' things and 'non-material' things. But new technologies are 'threatening' common distinctions and rusty ontological categories. Before computers and the internet, we might have categorized 'what is' into natural entities and artificial entities, but both kinds of entities were seen as material and physical. Some would also have added non-material and non-physical entities to their definition of 'what is' (for example, souls, ghosts, spirits, gods, daemons, angels, etc.), but these entities would not have been connected to the natural or material – let alone to *technology*. The picture would have looked roughly like this:

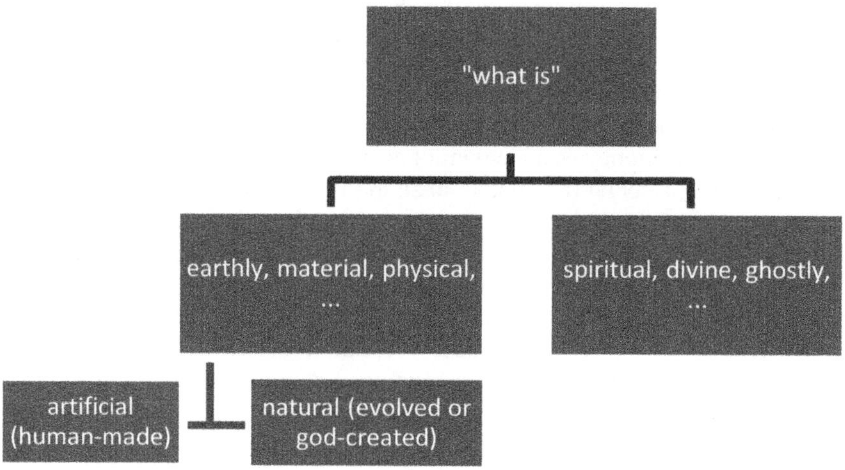


**Figure 5.1 Ontological categories**

However, after the 'digital' revolution or 'information' revolution, a new category of entities has emerged that is difficult to categorize within this scheme: 'digital' objects or 'virtual' objects. On the one hand, they are clearly human-made, and would normally fit the 'artificial' category. On the other hand, they do not easily fit the larger category of 'earthly, material, physical'. They might be 'earthly', perhaps, but they are non-material and non-physical. Perhaps we have to introduce a new distinction and add another category by splitting the category 'earthly' into 'material/physical' versus 'digital/virtual', with the latter subcategory only containing human-made things that are 'earthly' but neither material nor physical. Or should we rather link 'digital' and 'virtual' objects to the spiritual?

Whatever the 'correct' categorization may be, within object ontology reality is seen in terms of things, entities, objects. For ICTs, this means that within this way of thinking, the new reality created by the information revolution and the 'electronic' age is interpreted as consisting of new 'things' which we then try to include in the order of things and put in a (new or existing) category. For money, this means that it also appears as an object, a thing. But what kind of object? As we have seen in Chapters 2 and 3, it seems that the nature of money has changed. In the beginning of its history, money could be safely classified as a material thing, since it had the form of a valuable object or a coin. But a banknote or a piece of paper that is debt-related is already more difficult to categorize: it is material and human-made, but it already has acquired an immaterial quality. It is a thing but not *only* a material thing: its meaning and use reaches out into the immaterial sphere of meaning. 'What it is' cannot be reduced to its material or physical appearance. Perhaps even so-called 'material' monies such as coins always had an immaterial aura and function. (Maybe this is true for many things – natural and artificial – since humans ascribe meaning to things, they pick up more than just its material quality when they look at them and use them. I will say more about this below.) Thus, even if we consider money at an earlier stage of its development, we can raise doubts about its ontological status. At first sight it seems to be a 'material' thing, but on reflection it becomes much less clear in which ontological category money belongs.

In the electronic age, then, this categorization exercise becomes even more difficult as new forms of money and new financial entities emerge. Consider electronic money 'in' a bank account. The money that is 'in' or 'on' my electronic bank account can be turned into something more material (banknotes), but what 'is' it when it is on my account? And 'where' exactly is it? Moreover, what if a currency is entirely immaterial, such as Bitcoin or in-game currencies? What is its ontological status? What is 'virtual' money? Does it 'exist', and, if so, in what sense, exactly, does this money exist? It seems that there are 'virtual', 'digital' or 'electronic' things with no counterpart in the real world, although through exchange they can be transformed into a different, 'real' digital or electronic currency, which in turn can be materialized. (Cash machines function as 'materialization machines'.) Money thus seems to have 'magic' properties: apparently it can take on many different forms. Considering the 'metamorphosis' of money when it passes through various media, the speed of electronic communications, the 'omnipresence' of electronic money (in global electronic times it is everywhere and can be everywhere at the same time) and maybe also its 'omnipotence' (it is extremely powerful – see the discussion about power in the first chapters), money even seems to have spiritual, 'ghostly', or divine properties. But can anything that is human-made have such properties? Can anything 'earthly' have such properties? Again there is a categorization problem.

Similar problems are raised by numbers, another type of entities used in finance. What kind of ontological status do numbers have? Are they physical or not? Are they natural or artificial? Are they real or virtual? Are they 'digital'(!)?

According to Plato, numbers exist in a kind of abstract heaven, independent from our minds. Others say they are ideas in our minds. Or are they part of the real world? The information age challenges us to ask these questions again. Can we say that numbers are 'in' the computer or 'in' the network? If they are not, then 'where' are they? Are they 'on' or 'in' the screens we use? Are they in our mind? Can they be at several places simultaneously? Are they in data streams? But what is the ontological status of such 'flows'? There is, of course, a vast literature in philosophy of mathematics on the nature of numbers. All I want to say here is that object ontology asks these types of questions – all related to the main object-ontological question 'If it is a thing, what kind of thing is it?' – and that new (and old) financial technologies pose considerable challenges to this way of questioning and thinking. Are we perhaps asking the wrong kind of question(s)? Before moving on to different questions, let me first explore if there might be a solution within object ontology.

One way one could try to solve the categorization problems raised by money and especially electronic money is to accept that things can belong to different categories at the same time, that their ontological status is 'mixed' or 'multiple'. Money, and especially electronic money, can then be seen as belonging to several categories at the same time. It sometimes seems that it belongs to *all* categories (except the natural one perhaps). Bitcoin, for instance, has a 'digital', immaterial form and is 'earthly' and human-made, but it also seems to have 'spiritual' or even 'divine' properties since it is everywhere at the same time; in addition it can also become material and show its powers through other currencies. Note also that since bitcoins are generated by the computer, it is even questionable if it is merely 'human-made': the software is written by humans, but the production is done by the machine. Perhaps we have to add a category of 'computer-made' (or algorithm-made, robot-made, etc.). It turns out that Bitcoin is a good example of the multi-faceted nature of money in the electronic age. Similarly, one could see numbers as being both 'virtual' and 'material/physical': they are 'everywhere and nowhere' but they are also partly material/physical in the sense that they are produced by ICTs and they appear on material screens. Numbers also refer to quantities of material things. Maybe numbers are both part of the physical world and part of the human mind. Bitcoin also seems to have this dual virtual/material and mind/physical nature.

This approach seems to help, but one may wonder how useful these categories are if the new kind of entities have (nearly) all of them. Maybe in the information age we need new categories if this approach is to have interpretative force at all. Moreover, even this 'mixed' ontology starts from the same object-centred and dualistic thinking as the 'pure' ontologies mentioned before. The problems of object ontology to account for numbers and electronic money and currency are made possible by (1) the very idea that money and numbers are 'objects' or 'things' and by (2) a dualistic view of the world, which thinks according to dualistic categories such as real/virtual, digital/non-digital, physical/spiritual, etc. It is only if we believe that money is a 'thing' that we can ask questions such as

'What kind of "thing" is money?' and it is only if we assume that we can clearly distinguish between, for instance, 'digital' and 'non-digital' that we get into trouble categorizing the new 'things' created by new technologies. In the later sections of this chapter I will suggest other possible views, which enable us to move beyond the 'scholasticism' of the previous discussion.

Another way to try to solve the categorization problems mentioned previously, however, is to resort to what we could call 'mono-ontological' accounts. This approach is still family of the object-ontological approach, but avoids having to choose between different categories by claiming that there is just one category, that there is just *one kind of thing*. In the next section I will illustrate and explore such an approach, and show what it means for thinking about money and (electronic) financial technologies.

### *5.2.2. Information ontology: Money and numbers as information*

In order to avoid the specific object-ontological questions asked in the first section (e.g., is it digital, is it physical, etc.), one could say that all entities are basically, fundamentally just of one nature. We may think that there is a plurality of entities, but, so this argument goes, there really is just one type of entity. For instance, one could claim that money, numbers, and other financial entities are experienced as different phenomena, but that they are really *information*. Money and numbers in finance are information about what people are prepared to pay, how much they value particular goods. Numbers are information about price, about markets, etc. Let me further unpack this approach and its implications for thinking about money and its contemporary forms.

In the field of computer ethics, Floridi has proposed a philosophy of information that starts from an 'information ontology': an ontology which takes information as the primary ontological category. He argues that, at least at a basic, metaphysical level of abstraction, all entities are information or 'informational entities'; everything is fundamentally information. This re-conceptualization of ontology in informational terms is made possible by technological transformations. New ICTs invite us to 'consider the world as part of the infosphere' (Floridi 2013, p. 10). Floridi suggests that this is already happening today. He predicts that soon we will no longer think that there is a 'real', material world versus a 'virtual world'. Instead, there is only one world, the 'infosphere':

> The infosphere will not be a virtual environment supported by a genuinely 'material' world behind; rather, it will be the world itself that will be increasingly interpreted and understood informationally, as part of the infosphere. At the end of this shift, the infosphere will have moved from being a way to refer to the space of information to being synonymous with Being itself. (Floridi 2013, p. 10)

This means that categories such as 'digital' and 'analogue' are no longer relevant to describe ultimate reality, that is, they are no longer adequate at *that* level of abstraction. Floridi writes that 'digital and analogue are only 'modes of presentation' of Being (to paraphrase Kant)' (Floridi 2009, p. 152). He endorses the view that there is a 'mind-independent reality' and this reality is informational: 'as far as we can tell, the ultimate nature of reality is informational'; it is constituted by 'structural objects that are neither substantial nor material … but cohering clusters of *data*' (Floridi 2009, p. 154). We live in an informational environment, the 'infosphere'. This also means that humans are informational beings. We are part of an 'information ecology', and at one level of abstraction 'we can look at ourselves … as informational organisms, or *inforgs*' (Floridi 2008, p. 190).

For thinking about money, numbers, and other financial entities, this information-ontological approach implies that although at other levels of abstraction (and thus with other goals of analysis) we may say different things about money, numbers, etc., metaphysically speaking they are all informational entities. To say that entities are 'digital' or 'analogue' makes sense, this may be how we experience them at a given level of abstraction, but the ultimate nature of reality is informational. Applying this view to money, it means that although money may 'present' itself as 'digital', 'material', or otherwise (at various levels of presentation), from a fundamental, observer-independent perspective it is *information*. Through the lens of information ontology, there is no longer doubt about the ontological status of money and other financial entities: they are all basically informational entities, and the human persons who work with them are also basically informational. For instance, considered from an information-ontological perspective, traders at financial exchanges are financial *inforgs*; they are part of an information ecology that comprises all the financial entities under discussion, including electronic money, financial products, and the traders. And from this point of view it is true that regardless of how bitcoins present themselves to us and how they can be analysed at *other* levels of abstraction, their metaphysical nature is fundamentally informational; after all, Being itself is informational.

The advantages of this approach are clear: the confusions of (standard) object ontology are avoided, and it provides an elegant way of defining the nature of 'electronic' money: both 'digital' money and 'physical' money are information, even if they may 'present' themselves 'digitally' or 'materially'. Moreover, in its information-ecological version, this approach enables us to attend more to relations – hence the 'ecological' dimension of this kind of thinking. As Charles Ess has argued in *The Information Society*, the philosophy of information can be seen as a relational ontology given its emphasis on 'relationships between diverse elements' and 'the interconnection between all things' rather than on 'atoms', on 'distributed systems' rather than 'the single human being' (Ess 2009, p. 161). However, I see at least the following two problems.

First, although in principle this approach leaves room for looking at entities in different ways (that is, we can consider them at different levels of abstraction), by saying that 'ultimate' reality is informational it promotes one way of viewing entities over others (the informational one). It is thus more normative than it pretends to be. Its informational discourse tends to exclude other ways of viewing entities, in particular all the (other) ways the entity may 'present' itself. Although the approach explicitly acknowledges the existence of different modes of presentation and levels abstraction, by categorizing what is shown in these modes and at these levels as *mere* phenomena as opposed to ultimate (informational) reality, the approach in effect *marginalizes* them.

Second, considering its ontology and indeed its epistemology Floridi's Kantian gesture, and more generally any information ontology, remains fundamentally dualistic. On the one hand, there are the entities (e.g. informational entities), which are part of 'real' reality and are observer-independent, and, on the other hand, there is the observer, who remains out of sight. On the one hand, there are the phenomena (things may appear as 'digital' or not), and, on the other hand, there is the noumenal sphere which Floridi apparently has access to. What is missing, it turns out, is acknowledgment of the role of human subjectivity in relation to 'ultimate reality'. *Within* his information ontology, Floridi makes room for agents and patients, for looking at the world starting from different questions and (hence) from different levels of abstraction, but the very structures of reality remain untouched by human subjectivity. Viewed from an information-ontological approach, ultimate reality is observer-independent. Moreover, there is room for relations between things (after all, information ontology remains an object ontology or is at least close family of it) and in this weak sense there is 'relationality', but *social* relations are missing – they are not part of reality, unless they are framed in informational terms. This means that any analysis of financial technologies and media (such as new forms of money) conducted from within the information-ontological framework, cannot sufficiently account for the subjective and social dimension of these technologies since it does not have sufficient conceptual resources to do so.

At the end of this chapter I will propose a different approach that overcomes both problems and goes beyond dualism altogether by introducing an epistemological and social relationality that is 'stronger' or 'deeper' than the ones suggested by Floridi. But let me first introduce the social in a way that may be acceptable to dualistic and naturalist thinkers.

### *5.2.3. Social ontology: Money as social institution*

The previous ontologies assumed that the nature and meaning of money can be defined without the social, without social relations and social institutions, without society and culture. It was considered an object, independent of the social: an informational object, a material object, a digital object. But as we have seen in the chapters on the history of financial technologies and on distancing, money and other financial entities have always been deeply connected with social relations, society, and culture. For example, financial technologies had implications for social relations, and the emergence of particular financial technologies went hand in hand with the emergence of a particular type of society. The form of money also differed from one society to another. More generally, what something 'is' seems to depend on what (most) people say that it is, and we already share a language in which ontological distinctions are made, for instance between 'humans' and 'animals' – and therefore we share these distinctions. Acknowledging this social dimension to what (financial) 'things' are should have implications for our conceptualization of ontology, including the ontology of money. If we consider the status of a particular form of money such as coins, bank notes, 'electronic' money, Bitcoin, then this status always depends on what status it is *given* by humans and by society, on what status is *ascribed* to it by society, perhaps what status we agree on. For example, a piece of gold may be seen as very valuable in one society, but worthless in other. A bank note may be considered to be a mere piece of paper unless it is agreed that it has monetary value (e.g., exchange value). Numbers on the screen have no meaning unless, for instance, it is agreed that they represent monetary value (e.g., money on a bank account). And bitcoins are only bits, only 'code', unless we agree to assign value to it.

One approach to the metaphysics of money and other financial technologies, therefore, is to adopt a *social* ontology that takes into account the contribution of the social to 'what money is' or to 'what numbers are'. An example of a social ontology that does precisely that, and that remains in tune with the idea that (physical) reality is independent of the observer, is Searle's social ontology. Let me briefly discuss his view in order to shed further light on the metaphysics of money and other financial entities.

In *The Construction of Social Reality* (Searle 1995) and subsequent work (2006), Searle has tried to account for a type of entities that many people feel are in some way created by humans but at the same time are supposed to have an 'objective', observer-independent existence in the natural world. Searle's explanation of the nature of such entities and how they come into being crucially depends on the role of *language*. He explicitly gives an example from finance: paper money. There is a physical, natural reality, for instance a piece of paper, but this piece of paper *becomes money* when we collectively ascribe the status of money to it. Searle uses the term 'collective intentionality': we collectively accept that the piece of paper has a particular status, and therefore make it possible that the piece of paper performs a particular function. This collective intentionality thus makes the piece of paper into money rather than a mere piece of paper: 'the piece of paper in my hand ... performs a function not in virtue of its *physical* structure but in virtue of *collective attitudes*' (Searle 2006, p. 17). We collectively declare that this piece of paper has monetary value, we collectively give it a particular status. Searle calls this 'status function': we decide that 'X counts as Y in context C' (Searle 2006, p. 18).

This theory explains some of what I called the 'magical' properties of money, some of its 'metamorphic' characteristics. In particular, it explains how it is possible that not only a piece of paper, but also far less physical or non-physical entities, can have the status of money. Indeed in Searle's view, there can be status functions without physical objects (Searle 2006, p. 22). For the status of 'electronic' money and currencies (including Bitcoin), this means that, at least in Searle's view, it is not problematic that there is no physical object. What matters is the collective intention, the social construction of money. Financial entities are social institutions. Whether or not Bitcoin is 'physical' is not relevant to its status as currency, as money, as payment structure, etc. What matters is that it is becoming (has become, will become) an institutional fact. If and once there is collective intentionality, this is sufficient for something to have the function of money, whether it is digital or non-digital, electronic or non-electronic, material or non-material.

Moreover, the theory also explains the power of money, which now derives from the collective intentionality that supports it and created it in the first place. Once 'something' is given the function of money, this 'something' gives us the power to pay with it, to trade it, and so on. It may well be that there is only 'virtual' money in your Bitcoin Wallet or in your computer game; as long as there is collective intentionality which assigns the status of 'money' to it then it *is* money. 'What it is' depends on the social. It 'is' not only code, information, a virtual object, and so on; it is *also* what we, collectively, assign to it. In that sense, the currencies of the 'virtual' economy or the 'digital' money that is traded on Bitcoin platforms are as *real* as any other financial entity. Here 'real' does not only refer to the physical and material but also to the social. In contrast to the ontologies discussed in the previous sections, the social becomes part of the picture. This is a significant advantage in the light of the project to better understand contemporary forms of money and, more generally, contemporary financial technologies.

However, like Floridi's information ontology and (other) object ontologies, Searle's social ontology remains dualistic in the sense that there is a strict distinction between 'objective' reality and the 'subjective' world, between on the one hand the natural, physical (e.g., the piece of paper, which is physical and material) and on the other hand the social, cultural (collective intentionality, status functions, what language does and people do). Both Searle and Floridi start from a given, objective and 'hard' natural reality that then can be viewed at various levels of abstraction (Floridi 2004) or to which a 'social', institutional layer can be added by means of declaration (Searle 1995 and 2006). Now I would like to argue for a different approach, and explore its implications for thinking about financial entities such as (electronic) money.

### *5.2.4. Beyond ontology/Changing the question: A 'deep-relational' view of money*

Let me offer a relational view of money and other financial entities that goes beyond the ontologies discussed in the previous sections and, in a sense, beyond ontological thinking itself. From the point of view of object ontology, money is about objects, things, which are related to other things. Information ontology does not *fundamentally* alter this thinking: money is information (albeit perhaps of a specific kind), and informational objects are related to one another within an information ecology. Both ontologies assume what we could call a 'weak-relational' view: the focus is on the *relata*, not on the relations. There is also no need to talk about 'society' in this framework. There are relations between objects, and the objects may even be part of an ecology, but the focus is on matter, information, or whatever ultimate reality is held to be. Language can be used to describe this reality, but it does not really 'touch' it. Being is about entities, and Being can be known from an observer-independent point of view. There is also no need to talk about the social; the social is not relevant at that 'level of abstraction'. In sum, human subjectivity, language, and sociality are almost entirely absent from this picture. Searle's social ontology, by contrast, brings in the human and social dimension. It does so by adding a layer of humanity or sociality on top of observer-independent, natural reality. Language is used to bring new relations into being. Money becomes the material artefact plus whatever social status is produced by collective intentionality, by an act of declaration. Language and social relations enter the stage. However, it is still possible to sharply distinguish between the (non-human) physical-material and the human-social. At the end of the day, the question remains 'What is money?', 'What is Bitcoin?' etc., since in the Searlean approach human subjectivity only touches the surface of entities, the outer layer of social meaning. The material, natural core remains pure, untouched, and not transformed by human subjectivity and human sociality.

A truly different view, by contrast, one that involves a more radical relationality than Floridi's information ecology and takes the social more seriously than Searle does, changes the question from an ontological one to an epistemological one: 'What can we know?' It drops the interest in defining the ultimate reality or essence of entities (e.g. financial entities), and instead shows that the ontology game is *itself* one particular way of perceiving and constructing such entities. Even the term 'entities' already suggests an object-oriented perspective. The 'deeper', or if you like 'other', relation largely missed by the previous accounts is the relation between *object and subject*. To be fair, there are traces of it in Floridi's notion of levels of abstraction (we look at things differently depending on the purpose of our analysis) and in Searle's status function (the social status of something depends on how we construct it by means of declaration). But the 'nature' or deeper structure of things remains information (Floridi) or physical reality (Searle). Human subjectivity hovers 'above' (or, if you prefer, 'under' or 'next to') this 'real reality'. The object-subject relation remains out of sight (object ontology, information ontology) or is insufficiently problematized (Searle). A more radical-relational and more epistemological rather than ontological approach, by contrast, claims that to see reality in terms of information or physics is also and already a very specific way of looking at things, of constructing things. It is one that is motivated by the desire to dig deeper into things in order to uncover fundaments, to distillate from the phenomena their very essence – thereby assuming that there is such an essence and that we can have access to it. And the very game of ontology is one that takes distance from things and constructs them as non-human, one that opens up a gap between mind and mind-independent reality, between observer and observer-independent reality (matter, information, etc.). This renders it possible to see money as observer-independent (as object, as information) or as an assemblage of a part that is observer-independent and a part that is social and observer-dependent (money is then composed of an observer-independent object plus a declared status). But when we say something about the 'is' of objects, we can only do so *as subjects*. We do not have unmediated access to reality. Reality is always reality-for-us, as knowing and socially related subjects, living in a particular society and culture. Therefore, claims made within the discourse of object ontology or information ontology are also dependent on subjectivity and sociality. Then we cannot clearly distinguish the 'phenomena' from the 'noumena', as Floridi does in his Kantian moments. Then we have to change the game from 'ontology' to 'phenomenology'. Then we have to recognize that to view money as a material, digital, or informational object, indeed to write a 'metaphysics' of money, is only one particular way of looking at money, is only one perspective.

Indeed, this approach implies that what money 'is' depends on, and is shaped by, what we say it 'is', and this is not only a matter of status functions that leave 'physical reality' untouched; what this physical, informational, or any other 'real reality' is, is also a matter of description, ascription, and even prescription. In other words, it depends on what we human subjects do with language and with one another. It depends on language and social relations. If there is indeed this stronger relation between object and subject, then we cannot know 'what money is' apart from human perception, human language, and social construction. Then we cannot know the supposed neutral substrate the human world acts upon, since there is always subjective, linguistic, and social mediation. Then we have to define money as always already deeply human and social – a definition which brings us closer to what we can learn from the social sciences about money and other financial entities and technologies.

Let me develop this point by inquiring into the relation between money and technology, and by distinguishing between various meanings the sentence 'Money is a technology' can have.

## 5.3. Money and technology; money as technology

In the previous section I asked the ontological question(s) regarding money and other 'financial entities': 'What is money?', 'What is electronic money?', 'What is Bitcoin?' It turned out, however, that this particular philosophical approach and 'metaphysics' already involved a very specific perspective on reality and on money: one which tends to see money as a 'thing' and thereby diminishes or even totally neglects money's link to human subjectivity and social relations. In order to repair this, we need to find a way of thinking about money and its contemporary forms (e.g. Bitcoin) that is philosophically and analytically satisfactory, but that also opens up the discussion again to what the social sciences have to say about this topic (and therefore to the broader discussion offered in this book). Let me therefore do a reprise of the previous analysis, but this time starting from the relation between (electronic) money and 'technology'. In the previous chapters I suggested that money is a 'technology'. What does this mean?

Technology is usually viewed as a means to an end, as a tool used to create something (the artefact) or do something (help humans). What does this mean for money and electronic money? In the terms of object ontology, this creative aspect of technology means that in so far as money is artificial, is an 'artefact', it is created and processed by using specific tools. For instance, printing technology makes paper money, and cash machines deliver it and (help to) process it. Today, new forms of money are produced by new ICTs. Bitcoins are 'mined' (produced, created) by computer technology (hardware and the Bitcoin algorithm) and a Bitcoin Wallet is kept on a computer. In the terms of information ontology, it means that 'money-information' is created and processed by using *information technology*. For instance, electronic money is processed by computers in networks. In the terms of social ontology, it means that technology assists the performance made possible by collective intentionality and declaration, since it *materializes* the declaration. (As far as I know, this is not noted or recognized by Searle, but it could be added to his view.) For example, if we collectively agree, and declare, that this specific piece of paper is money, then printing technology *materializes* this: money is printed and the artefact is then a materialization of the collective intentionality and status function.[1]

However, one could also say – as I suggested before – that money just *is* a technology, that money is not only an 'artefact' produced by technology but that it is *itself* a technology. It is an object technology in so far as it does something with (other) objects. For example, it is a tool that enables me to exchange goods. It is an information and communication technology in so far it enables us to process information and communicate. And money is also a social tool in so far as it does things with people. Money is a (nearly universal?) 'mediator': it mediates between things and things, people and people, things and people. And with (my interpretation of) Simmel we could say that it is also a 'medium' in the sense of a *milieu*, an environment. Money is an 'in between', geographically and in our social relations, but it is also a medium 'in' which we live. Our culture is a 'money' culture, and what money 'is' and 'does' shapes, and is shaped by, our society and culture.

But if it functions as a technology in this way, is it still a 'mere' tool, a means to an end, or does it also *change the end*? If humans and things are 'touched' by the forms of money they create and use, do they remain the same? In the previous chapters I have argued, influenced by Simmel and by Heideggerian thinking about technology, that there are processes of distancing, alienation between people and things, and between people and people. I even wrote that money makes possible 'moral distance'. Now from the point of view of the ontologies discussed in the previous section, this is nonsense, meaningless. Object ontology could accept

1 Consider also objects used in marriage rituals: rings, for instance, are artefacts that materialize the social institution and declaration of marriage, that perhaps *help to perform it*.

causal relations, but lacks the conceptual resources that would enable it to reflect on how subjects (humans) are constituted by objects (money) and vice versa, and how there are deeper, epistemically and morally relevant relations between financial technologies and the humans and societies they co-shape (and are shaped by). Information ontology, at least if it stays at the level of abstraction at which everything appears as information (that is, in so far as it remains an information ontology), cannot conceive of these forms of distancing and alienation because considered from the 'ultimate' point of view of information ontology, humans are inforgs, are made of the same informational 'stuff' as (other) things. Moreover, at this level of abstraction society *does not even appear*, unless it is conceptualized as an information ecology or perhaps an 'information society' understood as a collection of inforgs. And since Searle's social ontology understands the social in terms of (collective) intention and declaration, it does not have room for the non-intended dimension of social institutions and for relations between artefacts/technologies and humans that are *not* agreed, intended etc. Similarly, Searle does not seem to have room for a more Heideggerian or (later) Wittgensteinian understanding of language. In Searle's view, language is a *tool* humans use to do things, collectively or individually. In a more Heideggerian understanding of language, by contrast, humans live 'in' language and language is given to them. This means, I think, that when we speak about money, financial technology, or anything else, we cannot simply 'decide' or 'agree' on their meaning. Their meaning also 'uses us', 'decides us'. Neither language nor technologies are entirely under control of human intention and social agreement. There are gaps between what we declare and what happens to and in our practices and institutions. There are gaps between on the one hand what we intend to do with technologies and with words, and on the other hand what happens to us, to our relations, to our world. What money 'is' and 'does', therefore, is not only a matter of social declaration and agreement. What money 'is' and 'does' will always partly escape our control.

A deep-relational view, therefore, attends to the mutual constitution of subject and object, and can conceptualize *that* and *how* financial technologies change humans and change social relations (and vice versa). Thus, whatever the 'ontological' status of money, electronic money, or Bitcoin 'is', these financial entities and technologies change what we 'are', what and how we do things, how we deal with one another, how we perceive reality, and in what kind of society we (will) live. If in the previous chapters I used the term 'financial technologies', therefore, this should not be understood as meaning 'financial tools' in a narrow sense: they are not only technologies in the sense of means to ends – with the ends being untouched by the means. They are not 'dead objects' separate from we thinking and living 'subjects', they are not only and perhaps not so much things intentionally used or constructed by (collective or individual) agents. They are technologies in a far broader and deeper sense: digital or not, electronic or not, virtual or not, informational or not, they transform what it means to be human and they change our societies. They transform our relation to the world, to others, and to ourselves. Dualistic and ontological thinking does not help us to understand these transformations, or to even see them. If we study and reflect on money and currencies as intrinsically human and social phenomena, we do a far better job.

Bitcoin, then, is not simply about 'digital objects' being manipulated by people and computers. It is not even merely a 'social construction' or 'social institution' if those terms refer to intentional (individual or collective) declarations and performances. Of course it is also about objects and it is also socially – and materially – constructed. But Bitcoin is part of a larger transformative process, in which all kinds of objects and subjects interact in various ways. It is about how we are changing, how our societies are changing. Financial technologies play a role in this process and are also created by it. It is a process that is at the same time material and dematerial, informational and non-informational. What matters is not the ontology. What matters is how we *experience* reality (this is the epistemological question, consider for instance the issue that experiential possibilities are made possible or discouraged by the technology, by society, etc.), what we do with the technologies and what the technologies do with us, and in which direction the process should go and what we can and should do about it (the normative issues, the ethics and politics broadly understood). For instance, do new forms of money and currency contribute to alienation from the local community and to impersonal relations between people? Do the same technologies perhaps open up different, alternative experiential and action possibilities? What can and should citizens do if they want to deal with societally relevant financial issues in a responsible way, either individually or collectively? In the next chapter I will say more about distancing and responsibility (in particular 'moral distance') in the light of contemporary global financial trade technologies.

## 5.4. Money and (other) social and political institutions; from trust in the nation state to different forms of trust

To end this chapter let me say more about the relation between money and institutions. Acknowledging the social and relational nature of money does not only mean recognizing that money itself has an institutional dimension or that it 'is' an institution, but also that it is related to various *other* social institutions (e.g., stock markets, private property, various political institutions) that it shapes and is shaped by. In the light of my discussion of financial technologies and globalization, for example, it is interesting to note that in line with Simmel (1907) and Searle (2005) most of us would connect money with the institution of the nation state. The nation state seems to provide the context in which something counts as money. In particular, nation states give legitimacy to specific forms of money; with their guarantees and indeed with their use of a particular form of *currency* they support or even create trust in 'their' currency (and/or in the currency of another state). For instance, the dollar counts as money in the context of the US state and in states that use or accept (exchanging) the dollar as currency. The euro counts as money in the context of the nation states that form part of the euro zone, and in the context of nation states that officially use the euro or accept (exchanging) it. Both currencies are already special in the sense that they are more 'global' than many other currencies, but in any case according to this view it is the nation state that guarantees its value and contributes to its acceptance, legitimation, and stability. (See also again the history of financial technologies.)

But if one accepts this theory, it seems hard to account for the use and (relative) success of Bitcoin as a currency, which is designed *not* to be dependent on institutional support from the nation state. More: so far it largely escapes control by the nation state, which is one reason why it is also used by criminals. Does this mean the currency is necessarily unstable and will remain unaccepted? Proponents of this view could point to the fluctuations in the exchange rates of Bitcoin, and to the fact that the currency is not widely accepted and used. However, I think there is also an alternative way of understanding Bitcoin and similar decentralized electronic currencies that are independent from state support. Bitcoin is also a social reality, it is not 'purely technical' as some may think. Now this social-relational nature or its context should not be confused with the nation state. It seems that, in principle, currencies can also function in *different* social contexts and be linked to *different* social institutions. Think about the semi-global currencies dollar and euro mentioned before. If someone in, say, a South American country accepts dollars, then this is not because her government supports the dollar in a formal way. That person may not be part of the US national institutional context at all. The status of the dollar depends on social acceptance (in this case, there need to be many people in that country who accept the dollar) but not directly on the institution of the nation state (only indirectly since the dollar is the currency of a country and government people seem to trust) and what happens to it is not even entirely dependent on intentions and agreements (e.g., the emergence of a crisis). Moreover, local groups and communities can create and use a currency that is linked to *their* social-institutional arrangements (even if that means they never totally control what happens to it). This can mean that within a particular territory or community both local and national currencies are used. With global electronic currencies such as Bitcoin this independence from national institutions is even stronger. Here the national context is entirely lacking. There is no 'reference' national context. Something (bits? code?) counts as money in a global context, or at least in the context of the community of Bitcoin users. This social or institutional context is important for the currency's 'life', but it is not a context that is anchored in a nation state. Bitcoin is itself become increasingly 'institutional' (it is a social-technological institution) and it is increasingly linked to other institutions (in some places you can pay with it, banks start to accept it, it can be exchanged in more places, etc.). This institutionalization does not happen because the nation state has sanctioned it, but rather because social acceptance grows (and because it starts leading 'its own life').

Furthermore, because of this decoupling from the nation state, the relations and trust connected to the currency have a more 'horizontal', decentralized structure. Whereas before people mainly used the currency because they (often implicitly) trusted the national authority, in a peer-to-peer network the currency is linked to trust between peers. Based on this 'horizontal' structure, it is in principle possible that the stability of the currency grows as well, depending on 'horizontal' trust and on what happens beyond people's intention (a crisis, an event, may destroy trust again). More generally, if (national) state/governmental support is not one of the necessary conditions for a money token or currency (see again Simmel), then this opens up more space for alternative forms of money – and more generally alternative financial technologies – that are independent of centralized power. It explains why global electronic currencies such as Bitcoin can become more widely accepted and can become more stable, but it also explains and makes possible *local* currencies and financial-economic systems, which also flourish without state support and seem to be based on similar 'horizontal', decentralized forms of trust. Perhaps such new local forms of money could help solve the distancing problems. I will say more about this in Chapter 8.

Chapter 6

# Money machines and moral distance: Financial ICTs, automation, and responsibility

## 6.1. Introduction

In this chapter I am not so much interested in the metaphysics of new electronic financial technologies; instead I focus on the *epistemology* of their use and the *moral* implications of that epistemology: implications for moral *responsibility*. As I remarked in my introduction, the use of new electronic technologies raises questions such as: 'What do traders and investors using these technologies *know* about the distant people and places which are (literally) *screen*ed off but which they nevertheless influence with their decisions and "on-line" actions?' and 'If contemporary global trade and exchange is highly dependent on ICTs, to the point that "machines take over" (see next), who is responsible when the machines make an error, when the algorithms are malfunctioning and create global chaos?' As I suggested in the previous chapter(s), however 'virtual' we may think global electronic 'flows' are, electronic ICTs in global finance have *real* consequences, and it is important to discuss the epistemic, moral, and social dimensions of these consequences.

My reflection in this chapter is triggered by what we may call 'the automation turn' (or perhaps the 'AI' turn) in contemporary finance and trade: today finance and trade is not only 'electronic' in the sense that computers, screens, and networks are used to mediate the trade; in exchanges all over the world the very trade actions that used to be done by human agents are now increasingly delegated to *artificial* agents. The key word in this development is 'speed'. Whereas some of the previous chapters were about 'placing' money and financial technologies, this discussion of the moral geography of global finance and its technologies starts with what at first sight appears to be a case of 'timing'. Is the development of what we may call 'financial AI' moving us towards a financial 'Singularity', towards a point in time at which we can no longer comprehend and control what is happening? Or is global finance already *past* that point? And what about time 'in' financial practices? What matters is *when* to sell. Of course time and space are related. The technological challenge is again to bridge physical *distance* by means of ICTs. On global financial markets there is a competition – some call it 'war' (Ross et al. 2012) – to trade faster than others and to hunt down profit opportunities: to buy and sell instantly, to predict behaviour, and to take advantage of very small price gaps. The tools of the trade ('weapons') are financial ICTs: algorithms, computers and their processing power, screens, and servers. They are used in order to bridge distances as fast as possible. Ironically, this obsession with time means that in the world of this so-called 'high-frequency trading' (see the next section) *physical space* gets important once again, since it now matters, for instance, *where* fibre-optic cables are located and how long they are, and *where* servers are placed: putting them close to an exchange's server gives a competitive advantage of microseconds. Moreover, as we will see in the next chapter, the flows of money we consider here are not quite as 'ghostly', de-placed, as one might think, but are connected to specific people, places, and material infrastructures. Let me first analyse the automation turn and the problems this raises for moral responsibility and distancing, in particular 'moral distance'. In the next chapter I will say more about place and matter.

First I say more about the automation turn and the ethical problems in contemporary finance. I will also position my argument and approach in the relevant literature, emphasizing again that neither ethics of finance nor philosophy of technology (ICTs, AI, robotics) has paid sufficient attention to new *financial* electronic technologies and media. In my analysis I use philosophical literature (ethics of finance, philosophy of technology, philosophical thinking about responsibility), but I also draw again on social-scientific work; this will build a bridge to the next chapter. In terms of technologies and specific financial practices, I will focus on financial practices mediated by contemporary ICTs and, within that area, on (1) the use of electronic trading platforms and screens and especially on (2) the practice of high-frequency trading (HFT). Then I will develop an argument about responsibility and moral distance in relation to these technologies.

## 6.2. 'The machines' are already here: HFT and the automation turn in contemporary global finance

To say that 'the machines are coming' is to announce an outdated state of affairs: 'they' are already here. And there is a sense in which they 'take over'. Although financial AIs usually receive far less attention in the press and in science-fiction movies than, for instance, autonomous humanoid robots or new chess computers, intelligent autonomous artificial agents are active in many domains, including finance. Consider again high-frequency trading (HFT): computers trade at high speed in order to compete with other computers. The machines takes over. But strictly speaking this is not even a 'replacement' of human trade; today these financial 'money machines' do something that humans could not do. The speed and volume of information processed is so high that human traders could not possibly do the same job. The technology used is also known as 'algorithms' or 'algos'. Computer programs automatically decide on timing, price, and quantity of the order. They initiate and execute orders without human intervention. They process the information they receive electronically before any human can possibly process the information they observe. Milliseconds or microseconds (some say nanoseconds) matter – giving new meaning to the saying *time is money*. Time is money and distance is money. There is a so-called 'race to zero'. Investment banks such as Goldman Sachs employ a lot of people working on algorithms that battle other algorithms. As in all contemporary battles, technology plays a major role. The game is to take advantage of very small variations in price. Others try to predict the behaviour of (other) traders, based on data gathering. Humans could never spot an opportunity to sell or buy in less than a millisecond, or process so many data. The players in these games are non-human: algorithms.

The economic and financial effects of this 'algorithmic' trading on global financial markets are still largely unknown. Many people in finance claim that high-frequency trading creates reduced spreads (reduced gaps between bid and offer price), improves liquidity and offers other advantages to markets, such as fewer intermediaries and lower transaction costs. Others stress the risks, or even point to ethical problems. Some say that the markets are more volatile, or that HFT games contribute to 'predatory' behaviour. And what happens if an algorithm gets out of control, for example, because of a software bug? Some propose banning HFT, or at least slowing it down. The so-called 'Flash Crash' of 6 May 2010 is sometimes attributed to HFT and seen as an example of how things can go wrong, although there is still controversy about the precise role of HFT in the crash. There is also the example of a trading firm, Knight Capital, which lost millions in half an hour as a result of a malfunctioning algorithm. Thus, although there is some controversy about the effects of HFT, it has become clear to many observers that the new technologies have changed the game, that they may create significant risk, and that there are problems concerning our knowledge of contemporary markets and how they are and might be shaped by the new financial-technological practice. In a much-quoted speech called 'The race to zero', Andrew Haldane (Bank of England) said:

> The Flash Crash was a near miss. It taught us something important, if uncomfortable, about our state of knowledge of modern financial markets. Not just that it was imperfect, but that these imperfections may magnify, sending systemic shockwaves. Technology allows us to thin-slice time. But thinner technological slices may make for fatter market tails. Flash Crashes, like car crashes, may be more severe the greater the velocity. (Haldane 2011, p. 19)

More generally, since its introduction around the turn of the century, the use of HFT has changed the face of trading, and we – citizens but also experts – know little about the financial and ethical implications of the new technological developments.

Furthermore, even outside HFT, electronic technologies play an important role. Trading has increasingly become 'electronic'. Today most traders at stock exchanges or futures exchanges do not trade face-to-face at a specific physical location (floor trading) where they communicate by means of shouting and hand signals (open outcry). Many have also abandoned telephone-based trading.

Instead, they use electronic, computerized trading platforms in order to place buy and sell orders. These systems process incoming market data at high speed and, again, reach high execution speeds using the internet. Traders use graphical user interfaces, several screens, and adapted keyboards with keys for specific tasks. An example is the famous Bloomberg Terminal,[1] a computer system connected to a service that enables traders to access real-time financial market data and place orders. It looks like a kind of 'trade cockpit'. This raises questions concerning how this interface, this electronic medium, impacts the trade. What kind of experience and knowledge does it give traders? By 'knowledge', I mean the knowledge traders have of other traders and investors,[2] but also the knowledge traders have of people affected by their trade (I will return to this issue below.) And in so far as there is an 'auto-pilot' (see again 'algo' trading), who is responsible?

This chapter will develop an argument about the exercise of responsibility in finance given the role of electronic technologies in its trade and investment practices. Although this argument needs further elaboration, it sets up a framework for thinking about moral responsibility in electronic trade and finance, and substantially contributes to the main argument of this book about distancing, in particular about 'moral distance'. By analysing the problem in terms of conditions for the exercise and ascription of responsibility, the chapter aims to contribute to the project of specifying what may render contemporary, ICT-mediated financial practices more responsible – assuming that this is our goal and assuming that this 'our' includes traders, that is, assuming good intentions. I argue that in order to act responsibly, people working in trade and finance need to know what they are doing (epistemic condition) and have control over what they are doing (control condition), but that meeting these conditions is rendered difficult given the role of electronic technologies in financial trade. These difficulties concern (1) the way ICT mediates information used by traders to take their decisions which creates an epistemic gap in terms of their knowledge about other traders and their knowledge about the consequences of their decision for others affected by their decision and (2) the fact that ICT makes it possible that traders hand over decisions to algorithms, which means they have less control over what they are doing and over what is happening. Moreover, I point out that citizens and those who represent them also face these two types of problems (as related to the epistemic condition and the control condition), and that these problems need to be dealt with if we want not only more responsible financial practice but also responsible innovation, regulation and policy in this area.

1 For those who wonder how Michael Bloomberg (mayor of New York City from 2002 until 2013) made his money, this is the answer.

2 With regard to the issue of knowledge about other traders it may be interesting to note that in 2008 the London Stock Exchange launched a trading platform developed by LSE, Baikal, which allows investors to trade large blocks of assets anonymously, thus creating a so-called 'dark pool' that is not open to the general investing public.

## 6.3. Ethics of finance and thinking about technology

In my introduction I observed that since the financial crisis of 2008, media attention has been paid to ethical issues in banking and finance. The financial sector has been criticized (e.g. bankers and banks), and politicians have been blamed for policies that are seen as destructive for solidarity, justice, and trust. The academic equivalents of these responses can be categorized into at least the following three levels and approaches in the field of ethics of finance. First, at the individual level ethics of finance tries to define the ethical duties and responsibilities of individual professionals working in the sector. This is sometimes translated into Codes of Ethics for the sector. Second, at the level of corporations (e.g. investment firms, banks) ethics of finance discusses the ethical quality of the behaviour of these corporate agents, for example tensions between the aim to make money (making shareholders happy) versus responsibilities to clients and to other stakeholders. For example, it is generally seen as morally wrong to take advantage of the ignorance of 'naïve' small clients. Third, at the level of the wider society, questions are asked regarding the value and fairness of the social-economic system (e.g., 'capitalism' – some would say 'casino capitalism') and the desirability of the dominant political ideologies that currently inform the policies that govern the sector.

While these discussions have delivered some valuable insights into the ethical aspects of finance (see, for example, Boatright, 2010, 1999; Dobson, 1997; Heath, 2010; Kolb, 2010a, 2010b: Reynolds, 2011), it is striking that in the literature little attention is paid to *technologies* used in financial practices and their ethical consequences. Moreover, so far little work on this theme has been done by people working in ethics of information technology and philosophy of technology, including philosophy and ethics of (new) ICTs, AI, and robotics and automation. This is to be regretted since those fields can contribute to identifying, understanding, and coping with problems in ethics of finance. These disciplines could also expand their range, further develop their thinking about morality and ICTs, and make themselves more relevant to current global societal problems if they paid more attention to the field of finance and its technological practices.

An exception is a relatively recent article in the area of engineering ethics that surveys problems in high-frequency trading and proposes quality management techniques as a foundation for ethical standards in this area (Davis et al. 2013). An older but also still relevant and useful article by Hurlburt et al. (2009) analyses the financial crises of 2008 and argues that automation contributed and still contributes to market uncertainty, especially as risk management became automated. The authors claim that automation made possible a cavalier approach to (applied) risk management. In response to this literature, I will refer to specific arguments made by these authors and I will clarify the distinctive approach and contribution of my chapter. But these articles remain an exception and they are not centrally situated within the fields and discourses of ethics of information technology or philosophy of technology. They also lack a firm connection to thinking in moral philosophy. This chapter seeks to put the topic 'ethics of automation in finance' higher on the agenda of ethics of information technology (including AI and robotics) and makes that connection with moral philosophy.

In particular, philosophical reflection on responsibility is really helpful to enhance the public discussion about ethics of finance. It is to be regretted, for instance, that many commentators in the media tend to restrict the ethically relevant part of their comments to blaming particular individuals, classes of individuals ('investment bankers'), and corporations. I endorse calls for more responsible action in the sector. But academic reflection asks more from us. We need to go beyond easy questions and easy answers ('Are they bad?'/'Yes'./'Should they act more responsibly?'/'Of course they should'.). In this chapter I ask the question: assuming the people in the sector want to exercise their moral responsibility (and assuming that other stakeholders want them to do that), then *how* can they do this and how can this be *supported*? And, especially, how can they exercise their moral responsibility given the technological changes to their practices?

There are various possible answers and approaches here. For example, one could offer Codes of Ethics to the profession, and adapt those codes to the new contexts and indeed to the new technologies involved. This should involve asking broader questions such as: What does it mean to act responsibly in the age of electronic trading? How should traders and other parties involved understand their responsibilities? What values do traders have, and do these values change in the new technologically mediated contexts? One could also discuss the institutional, organizational and economic and societal context within which these people work and its relation to moral responsibility. For instance, one could ask the question whether the individual level of analysis is sufficient, given that both individuals and the automation technologies are part of larger organizations and contexts, i.e. firms, the exchanges, the regulatory authorities, the HFT industry. Davis et al., for instance, rightly emphasize the *organizational* responsibilities to external stakeholders (they say that what really matters is the behaviour of organizations) and point to the fact that people who develop and use the technology work in interdisciplinary teams of professionals. The teams include so-called 'quants' (people who provide the mathematics, who do the quantitative analysis) and computer engineers. These professionals all have their own perceptions and biases (Davis et al. 2013, p. 860) and their own 'ethical climate' (863), that is, their own aims, values, and interests. In addition, one could call for ethical principles that could guide the project of making the institutions in which these people and these technologies are embedded more ethical and more responsive to social concerns. There is micro-ethics and macro-ethics (see also Hurlburt et al. 2009) and I would also add: meso-ethics (ethics at the level of the organization).

This chapter's focus is on the relation between (automation) technology, knowledge, and responsibility. The emphasis is on a philosophical analysis of responsibility and its implications for ethics of financial technologies, not on the question of which ethical principles or ethical standards should guide the practice, which 'standards of conduct' people should follow, let alone which quality standards should be implemented in the industry (Davis et al., p. 862) – however useful and meaningful such discussions and efforts may be. Also, let me emphasize again that I do not assume that individuals working in the sector are necessarily self-interested. Here I am interested in *how* people *can* act in more responsible ways given specific financial technologies, and how financial technologies and policies can be designed that give people more possibilities for exercising responsibility. The aim is to explore how we can make possible better financial practices, thereby charitably assuming that practitioners want to promote their own interests and those of their organization, but also are committed to do this in an ethical way. Moreover, the analysis is presented in terms of individual expert responsibility, but the argument could be easily expanded to collective agents such as professional teams and organizations.

Having clarified my agenda, approach and assumptions, let me start with the argument. I propose to analyse the problem regarding the moral responsibility of financial experts and relate this problem to issues of distance (in particular 'moral distance') by relying on two traditional Aristotelian conditions for holding someone responsible, that is, two conditions for moral responsibility ascription.

## 6.4. An argument about responsibility and moral distance

In the *Nicomachean Ethics* Aristotle distinguished between two conditions for holding someone responsible: one must not be ignorant of what one is doing and one should not be forced to do something (Book III, 1109b30–1111b5). Fulfilling these conditions becomes problematic in the case of financial practices mediated by current ICTs for at least the following reasons.

### *6.4.1. The epistemic condition*

First, it is difficult for traders to know what exactly they are doing if (1) they lack direct contact to those they trade with (other traders) and if (2) they cannot overview the consequences of their actions for others. They lack direct and personal contact with other traders since those traders are situated at distant locations. Sitting behind the screens of their terminals, they try to interpret the market – and thus, indirectly, the behaviour of other traders – but the data they get are impersonal. As work in social studies of finance shows, there is a significant epistemic difference between the experience of the trading pit and current electronic trading. Direct contact by means of hand signalling and shouting is replaced by interaction with the electronic interface. There are data on the screens (there are several displays), there is a keyboard (specially designed for this type of trading). Zaloom has explained that whereas the pit involved 'full-body experiences', screen-based technologies give the trader numbers (Zaloom 2003, p. 263). This does not only require that the traders develop different skills in order to communicate; it also creates distance between the traders. Although they have still some contact with colleagues at the local trading floor (which becomes an *office*), generally their experience of trading is situated within a global space. Knorr Cetina and Bruegger have shown that traders have this double orientation, but 'the screen – and through the screen, the global sphere of transactions – is what is dominant' (Knorr Cetina and Bruegger 2002a, 923). I suggest that having this 'double' experience is not unlike the work experience many of us have in the office (or *other* offices). Consider academia. Due to the internet, our primary orientation is now no longer the 'local' space and the people in the office or building. Most of us interact more with other people: people 'somewhere' on the net, somewhere in the global internet-mediated space in which we live. It also means that, in principle, work can be done from any location, provided we have access to the technology. Traders can now trade at home using their smartphone, for instance.[3]

While this move to electronic trading and electronic working does not make our experience less dependent on material technologies (i.e., computers, wires, servers, etc. – see the next chapter), it is undeniable that the technology makes possible significant epistemic distance between us and our direct social and material (work) environment. Of course 'local' interactions are not excluded; we still have them (see also next chapter). The point of the 'distancing' as 'globalization' thesis (see Chapter 4) is that these local interactions become less important when most of our work is mediated by ICTs, and in particular that they play a less important *epistemic*, *social* and *moral* role. The primary epistemic, social, and moral environment becomes a global one. In many cases this also means less personal and certainly less direct, physical communication. Partly this globalization of work space also means that there is a danger that spaces become *de-socialized*. Due to quantification processes (see also Simmel), the world we encounter increasingly appears to us as a world of information, data, or numbers; it is no longer mainly a world of people. (Incidentally, this is also why information ontology or object ontology appear to be adequate descriptions of our world.) For trading, the lack of sufficient personal and physical communication with other traders means that the knowledge they have about their actions is about numbers and data on screens, not about what they do with or to people. Work is then experienced as non-social, a-social. At the same time, there are also indications that in global trade there *are* social spaces, some of which are very competitive and in which *anti*-social, 'predatory' behaviour is encouraged (consider again what is said about HFT). Either way, the *technologies* contribute to the emergence of these new types of spaces and new kinds of experiences and knowledge. Technology changes the practice and the space and thereby its epistemic, social, and – as I will argue – *moral* character.

'Knowledge' is also an issue in the sense that it becomes increasingly difficult to understand what goes on in such a global work space. When the financial and economic *agora*[4] changes from the local market place and shop to the 'global

---

3 This phenomenon needs further analysis. Here I will focus on (non-mobile) trading platforms.

4 In ancient Greek city-states the *agora* was an open space where people gathered, e.g. for markets and political activities.

agora', the global anonymous 'market' and 'markets' (and to some extent this already happened in ancient times), trade becomes more abstract and more complex. This is certainly the case in contemporary times, when global finance involves the new ICTs under discussion (HFT as automation technology) and indeed complex financial products: the objects of trade become so abstract that only computers can (literally) deal with them. To know what 'the market' does and to conduct their own trade actions, traders are entirely dependent on them.

Due to increased complexity and epistemic opacity there are also knowledge problems at the level of professional teams. Complex and abstract trade requires specialists. Since electronic automated trade involves professionals with different backgrounds (see also Davis et al.), this causes problems related to (mis) understanding what each is doing and how each other's tools work. For instance, Hurlburt et al. have argued that there are not only problems with the models made by 'quants' but that (risk) managers also fail to understand those models. I think this issue can be conceptualized as a problem of compartmentalization of knowledge, which is related to labour specialization: due to the difficult, abstract and complex nature of the work different kinds of specialists are needed, and specialization tends to create different 'epistemic boxes', different knowledge spheres or compartments, which then need to be connected. As in academia, much time then goes into getting people of different disciplines to talk to one another (in a way that is helpful in order to tackle the problem at hand). Different epistemic worlds means different languages, and connecting them requires translation (which is extra work). And there are also other ethical problems here, such as the fact that the programs tend to be biased to provide value to shareholders (Hurlburt et al. 2009). If the technology is used to promote short-term shareholder value, then there are a lot of stakeholders that remain out of sight.

But this is also related to a *moral* problem. How is it possible to act responsibly if you do not even *know* the consequences for non-shareholders? It is plausible to suppose that traders in contemporary global finance do not know much about the consequences of their actions for people whose lives depend on the price of the financial products (stocks, futures, etc.) that are traded and, more generally, on their financial (trade and investment) decisions. In spite of their 'global' epistemic orientation, traders lack sufficient insight into the precise consequences of their trading actions for people living in their own country and in other countries in the world. Metaphorically speaking (and sometimes literally), the 'financial global city' becomes an island that 'drifts off' or 'lifts off' from 'the rest of the city', 'the rest of the world'. Traders 'on the island' or 'in the tower' do not experience what it means for people 'over there', 'on the ground', if for example the price of wheat goes down or if the price of oil goes up. Traders make money for their clients and for their (investment) firm (and indeed for themselves), but *as traders* they are largely ignorant about its global and local impact on the lives of people. Journalists may try to reveal links between financial decisions and 'the real world' (e.g., the people affected), but there is no guarantee that traders receive this information. On the 'island' and sitting in their electronic 'cockpit' (see Chapter

7), they are likely to have only very general ideas about what and whom they are trading with and about the effects of their actions on 'the market'. Given the complexities of contemporary global economics and finance it is not even guaranteed that *anyone* has a good overview of the relevant causal relations. There may be some information available 'out there' (e.g., information about the relation between derivatives and food prices) but the knowledge researchers have is still rather limited – let alone that specific information reaches the relevant traders in their 'cockpit' or the relevant managers in their lofty offices. We could compare the epistemic situation of the (HFT) trader with driving a car on public roads at high speed while seeing only 20 metres in front of you because of the fog or because you have 'tunnel vision'. Watching the dashboard does not help you to know if you are going to hit a pedestrian – or indeed anything or anyone out there. The only thing you know is that you're going fast, perhaps faster than the other racers. What happens to the market(s) and to the people who are not directly involved in the trade actions remains largely unknown. The speed of the technology and the interface create too much epistemic distance and hence 'moral distance'.

Note that the HFT technologies and the electronic interfaces do not completely rule out human judgement, at least in so far as humans are still 'on the loop', meaning that they still supervise the technology; the scope for human judgement depends on the degree of control. I will say more about the problem of control below. But with regard to knowledge it is important to remark (especially keeping in mind traditional technology criticism in philosophy of technology and also Simmel's criticism) that the kind of thinking involved in trading itself is not purely 'calculating'. As Beunza and Stark have argued, traders calculate but this calculation is 'far from mechanical' and 'involves judgment' (Beunza and Stark, 2004, p. 371). This implies, I think, that traders (and the 'quants') cannot just evade responsibility for their actions. The 'machine' metaphor is not entirely suitable for describing the situation if by 'machine' we mean industrial-type mechanical systems from which there is no escape and in which we are mere cogs (or indeed the product being made), machines without any 'human' element let alone the possibility of judgement. But in trade often 'judgement' is limited to the financial decision itself and its goal (financial profit for the firm and for oneself). If traders suffer from a kind of moral-epistemic 'tunnel vision', then the task is to widen the field of judgement beyond the narrow space of trade decisions to include the moral and social consequences of their actions. Otherwise there is again too much 'moral distance'.

Note also again that these epistemic problems related to the technology come in addition to epistemic problems with financial products. Crotty (2009) argues that the global financial crisis is rooted in structural flaws in the financial system, which include failing regulations and incentives that increase excessive risk, but also innovation that created very 'complex and opaque' financial products (p. 566). If we create such products, then *of course* we need the new machines. Then we have to delegate our actions to the computer. But why do we want these types of financial products, if it leads to distancing and de-humanization of the practice?

We seem to be on a rollercoaster ride fuelled by a combination of financial and technological innovation, and nobody knows where we are going, or even what we are doing.

### *6.4.2. The control condition*

The 'rollercoaster' metaphor also brings us to the second Aristotelian condition of responsibility: control. How much control do traders have over their actions, and, as I will ask in the next section, how much control do we citizens still have over what is happening in global finance? When trade is increasingly automated, there is considerably less room for exercising responsibility. How can someone exercise responsibility if (s)he is no longer in charge? Although today humans are still 'in the loop' or at least 'on the loop' (to use terms from discussions about military automation technology), and although the amount of control is a matter of degree (it is not an all or nothing question: surely traders have *some* control), this does not render the control question less relevant. If Aristotle is right about control as a condition of responsibility, then we must ask: is it possible for traders to act responsibly if and in so far as computer programs take over their trade decisions?

For sure, asking the question as if it is either humans or technologies that do the work is a little misleading. Perhaps it is true that, as social studies of finance show, cognition is anyway *distributed* among networks of tools including 'computer programs, screens, dials, robots, telephones, mirrors, cable connections, etc.' (Beunza and Stark, 2004, 389). The same could be said about trade actions: they are also distributed among such networks. Humans and technologies combine to bring about the trade (see also the next chapter). But acknowledging this distributed character of trade practices does not render the question concerning control irrelevant. Before electronic trading, humans were at least much *more* in control, even if they also used certain tools and machines. Now, as the 'information revolution' accelerates and has also transformed global finance, they seem much *less* in control. The distribution has changed. In automated trading, for example in HFT, the computer *takes over* at least in the sense that the tasks required from a trader today can no longer be performed by humans alone. Of course there are still humans involved in the sense that, as a director of a HFT company has pointed out in *The Financial Times*, 'people design these systems and strategies, programme the computers, supervise the trading and manage the risks' (Lenterman 2013). But just as the human pilot of a contemporary aeroplane hands over the plane to the autopilot (while maintaining supervision control, while being at least 'on the loop' as a flight manager), so too high-frequency traders delegate trading to *their* 'autopilot'. Thus, the cognition is not only distributed, but the cognition and the *action* is to a significant degree *delegated* to machines, even if those machines and computer programs are made and supervised by humans.

This type of situation raises questions known to the field of computer ethics and philosophy of artificial intelligence (e.g., 'machine ethics') as having to do with *the status and responsibility of autonomous artificial agents and robots* (see, for example, Wallach and Allen 2009; Floridi and Sanders 2004; Himma 2009; Stahl 2004). If an artificial agent, say an 'algo', makes a mistake, who or what is responsible for this mistake? Can 'algorithms' be regarded as *moral* agents? Can they be held responsible? If not, who exactly *is* responsible? The programmer, or the investment corporation? Should responsibility also be distributed? Should it be distributed among humans and non-humans? And should we develop *moral machines*, by giving them a kind of built-in 'morality'? Should we make the 'algos' more ethical? *Can* we make them more ethical? What kind of 'morality' would such artificial agents have, if any? Will the further development of artificially intelligent financial agents lead to what thinkers in the field of AI sometimes refer to as 'The Singularity'? Will machines 'take over'? What is and should be the place of the human in finance?

Although a comprehensive discussion of all these questions is beyond the scope of this chapter, it is clear that the control condition is difficult to fulfil if and *to the extent that* trade decisions are delegated to machines. If an algorithm is malfunctioning (or 'misbehaving') and it takes people who run it half an hour or more to regain control (perhaps even by switching off power), then clearly humans are *not* fully in control. If people are using algorithms but cannot do much about what happens to the market (e.g., in case of a 'flash crash'), then humans are *not* fully in control. This means that even if people *would like* or *intend* to act responsibly (and, as I said, this is what I charitably assume in this book), it is *difficult* to do so given the situations and practices shaped by the new technologies (and of course also by the humans designing, employing, operating, and supervising them). Thus, from the point of view of prevention, we have a problem. But we also have a problem when we want to deal with 'accidents' or 'catastrophes' that already happened. With automation, there is a danger that in cases when something goes wrong (and it went wrong and it will go wrong again in the future), traders deny and evade responsibility – 'it's the algorithm!' – or that in any case the ascription of responsibility becomes really difficult. If action is distributed between humans and machines, this is always the case. Think about plane crashes: did the autopilot make an error, or did the human pilot make a mistake? Did both of them make a mistake? Long reports and entire books have been written about such problems with planes and space technology; and 'crashes' and other 'disasters' and 'catastrophes' related to new financial technologies raise and *will* raise similar issues. Automation renders it more difficult to sort out who is responsible for what to what degree (if we ask the question from a third person perspective, that of the observer) and it renders it more difficult to understand one's own responsibility as an engineer, manager, quant, etc. – that is, *my* responsibility is at stake here rather than the responsibility of someone else (if we ask the question from a first person perspective, that of the trader involved). For professionals, it is important to understand that ethics or responsibility is not (only) someone else's job. It is not the responsibility of 'one', to use a famous Heideggerian term, but it is about *me*. In so far as the use of financial ICTs, in particular automation technologies, contributes to diluting responsibility and to increasing the control distance between humans and the (trade) action, exercising this personal-professional responsibility is rendered more difficult. In this sense, then, financial automation technologies contribute to increased 'moral distance'.

### *6.4.3. Conclusions for the responsibility of professionals and citizens*

Given these problems, I conclude that if we want professionals in the financial world to act responsibly, we should further investigate the relationship between financial technologies and responsibility, and think about how financial-technological practices can be made more ethical, which means here: think about how the epistemic and control gaps can be bridged, how we can cope with the 'moral distance' problem. Again, the main challenge is not who to blame for what goes wrong, but how to make it possible that the professionals involved *can* exercise moral responsibility; we have to improve the conditions under which they can do this, assuming that they *want* to and *should*. From a public and societal point of view, responding to this challenge may include intervening in the regulation of financial practices (or lack thereof), but also encouraging the design of new *technologies* and hence new technological practices. We do not need a kind of revolt against technology as such; perhaps not even against automation technology. Instead, we need to create technologies and social arrangements and institutions that make possible *different* epistemic relations and control relations that are less 'distant'. If it is true that technologies – in particular ICTs – play such an important role in current global financial practices, then responsible research and innovation of this area must be a key element in any viable strategy for dealing with the 'moral distance' problem. This means that not only traders but also 'quants', engineers, and other people involved in the development of new financial technologies need to reflect on *their* responsibility, and that policy at national and supra-national level should promote responsible research and innovation in these areas. The same is true for (other) financial experts who design new financial products and for managers, investors, and bankers directly or indirectly involved in the trade: their responsibility and the ethical character of their research, innovations, procedures, and decisions are also at stake.

Furthermore, the 'me' and the 'I' at the end of the previous section includes not only the trader, the manager, the programmer, or the investor. We should target the responsibility of financial experts and financial actors, but also ask the question how at societal level *all of us*, as citizens, can take responsibility for what happens and make things better. At the societal and political level, we find similar moral-epistemic problems, which can be identified and ordered according to the same Aristotelian conditions for responsibility ascription.

First, can citizens or politicians act responsibly in relation to global finance if they do not really know what goes on in contemporary techno-finance, if they do not really understand the financial sector and its technologies? Politicians, civil servants, and (other) regulators – let alone other citizens – do not always understand electronic trading. As I suggested, not even the managers directly involved fully understand it. How can citizens and those who claim to represent them then make laws and regulations, create incentives, install proper monitoring procedures, and build in adequate safeguards, if they lack knowledge of the financial world and its practices and technologies? There seems to be a significant epistemic and hence moral distance here, which threatens the proper working of our democratic institutions. We must reflect on how financial experts, but also journalists and citizens, can help to bridge that distance and contribute to more and better (shared) financial knowledge, more transparent and responsible financial institutions, and indeed more transparent financial products, technologies, and practices. This may include what we could call 'bottom-up' techno-social solutions and experiments. (In Chapter 8 I will explore alternative financial technologies.) If 'they' do not solve the problem (because 'they' cannot or because they do not want to), then we (individuals, 'civil society') could try to do better.

Second, do we, as a society, have sufficient control over global finance? Many people in the financial sector claim the sector can and must regulate itself and many politicians seem to follow 'the markets'. But can we leave moral, political, and technological change to the financial sector and to 'the markets'? Would this not amount to evading responsibility? *Can* we citizens (re)gain control, given developments in financial automation, which seem to take place in a global context? And if the space for political-financial control and responsibility is limited, then can we change technological-social and political arrangements in such a way that this space is broadened? Again we observe a 'control' type of distance that is morally and politically relevant, and we can explore and experiment with new technologies and also new policies and new political arrangements that aim to bridge that distance. For instance, the nation state and its institutions are not necessarily the best framework for dealing with the problems identified; we can at least explore democratic alternatives at supra-national and sub-national level. And as suggested the development of alternative financial technologies and social arrangements may also come from 'bottom-up' rather than only 'top-down' (see again Chapter 8). We may blame 'them' for our lack of power when it comes to 'global' finance; but perhaps we have more power than we think – or we can (re) claim some of it – and use it to contribute to processes of techno-social innovation.

In sum, the epistemic gap and the control gap identified with regard to financial professionals and their technologies are similar to gaps in our democratic fabric as it has been transformed by processes of globalization and by the new technologies of the 'information revolution'. Politicians and other 'public' professionals have an important role to play when it comes to dealing with these gaps. But citizens have their share of the problem and can also contribute to coping with it. If we want to act responsibly as citizens, then it is our democratic duty to try to better understand the financial world and to address the call for responsible action and innovation not only to people working in that world (to 'they' and 'them') but also to ourselves as citizens.

## 6.5. Conclusion

So far, contemporary financial technologies have received little attention within computer ethics (ethics of information technologies) and philosophy of technology. This is unfortunate given the financial and economic challenges we face as societies in the context of globalization, and given the social and political, democratic, implications of ethical problems in finance. The world of finance is rapidly changing due to new ICTs and this deserves our urgent attention since it matters to the lives of many people, that is, the lives of people who are usually not directly involved in electronic trading or in the 'financial sector'. It matters to *our* lives. Financial ICTs in the context of globalization have made possible a highly interdependent world in which financial practices in one part of the world have consequences for people in distant countries whose lives and work depend on prices of goods. It is also a world in which technological change, for instance in the area of high-frequency trading, has consequences for global finance at large, and thereby for all of us. By discussing the relation between financial technologies and moral responsibility this chapter has shown how we can begin to do conceptual work in this area in a way that benefits from work in social studies of finance and makes connections with ongoing work in computer ethics and philosophy of technology, especially in the area of ethics of robotics, artificial intelligence, and related areas.

For philosophy of technology, 'machine ethics', and ethics of AI and robotics, this discussion suggests that we need to direct our normative philosophical efforts not only to by now familiar areas such as social media, military robotics, ICT in health care, etc., but also to 'machines' in *finance*. 'Machine ethics' is not only about artificial intelligence in a vague, distant *future*, about algorithms that are at work when we search the internet, or about highly visible *robots* we might encounter in the hospital and in the home; it is also about money and about the machines currently at work in global trade and finance. That world is far less sexy and far less visible to most of us, but what is happening there in terms of technological innovation deserves attention from the 'machine ethics' community.

Moreover, as is always the case for ethics that aims to be relevant to societal problems, our thinking about these matters needs to be empirically informed. In order to fully address the problem of moral responsibility and 'moral distance' raised here, we need to know more about the concrete technological practices of contemporary finance and the 'knowledge' and 'control' issues that arise in these practices. In the next chapter, I will refine and partly revise my argument about financial technologies and distancing by further engaging with more empirically oriented work from other disciplines.

# Chapter 7
# Geography 2: Placing, materializing, humanizing, and personalizing global finance

## 7.1. Introduction

In the previous chapters I have argued how under conditions of modernity and globalization, financial technologies have helped to create a space of financial flows that is increasingly global, immaterial, and out of control. This has moral-spatial, social-spatial, and epistemic-spatial consequences: different types of distancing have emerged. It seems that we have been alienated from one another and from reality. Financial and other ICTs and digital media have shrunk the world, but, so it seems, only in a physical sense. It appears that we did not become closer to distant others, and both the moral relation and the knowledge relation we have to what happens elsewhere is of a detached, disengaged kind. I have also explored the implications of these forms of distancing for responsibility: is it still possible to ascribe responsibility in finance, or to exercise responsibility in finance, under these conditions? In particular, is this still possible if the screens of global finance give us too little knowledge about the ethical and social consequences of financial actions, and if its algorithms take over trading actions? The new financial technologies seem to have contributed to 'moral distance'.

In this chapter I nuance this picture of global finance and the impact of financial ICTs. Although there are indeed processes of distancing that follow the 'bridging' or 'shrinking' of distance made possible by contemporary electronic ICTs and digital media used in finance and elsewhere, there are at least a number of important 'footnotes' to be added to this story – if not 'revisions'. I will show that while it is undeniable that in so far as we are living in a modern technological culture and a globalized and globalizing world we are subject to processes of bridging and distancing, global finance is more place-bound, material, and 'human' than previously presented, and leaves more space for the personal than assumed by Simmel. It will turn out that the geography of contemporary financial experience and action is less global than one would expect (Chapter 4), that financial practices are less 'virtual' and more material and embodied than supposed in my reading of Simmel (Chapter 3) and my inquiry into the metaphysics of contemporary money (Chapter 5), and that in global ICTs-mediated finance as a world of 'money machines' (Chapter 6) there is room for personal relations and for judgement and interpretation – in other words, for the human. Financial practices have always been as much human as technological, and as humans we can take distance from our practices. As I will argue in the next chapter, this also opens up possibilities for financial-social change: for resistance and for alternative practices and technologies.

## 7.2. How 'global' is global finance?

### *7.2.1. Lessons from the discussion about globalization*

Contemporary electronic ICTs and digital media make it possible to interconnect people all over the planet, creating what might seem at first sight a 'global village'. But how 'globalized' is our economy and culture really, and, if it is globalized at all, does this necessarily mean that the local becomes less important, that place becomes less important? How 'global' is 'global finance', and is it really independent of places – hovering, as it were, 'above' them without any connections or anchors in local spaces and meanings? Is it true that, as Simmel argued, getting closer to the distant necessarily means getting more distant from what is close? Does globalization necessarily alienate us from people and places near to us?

First, while there clearly is a process of globalization, it has not (yet?) reached its full potential. The development of a truly global space of flows, including information flows and monetary flows, is still hindered by many barriers and borders. National governments often try to keep or regain control over what happens within its physical territory, for instance by trying to control the internet and by regulating flows of money. Legal frameworks are often national. Commercial actors such as banks use the persistence of territorial borders to make profit, for instance by asking customers to pay for international bank transfers. And many other 'flows' such as labour flows are still very territorialized, as anyone who has ever migrated to another country will have experienced.

There are many more limits to financial and economic globalization. For instance, territorially defined nation states still play an important role in investment and corporate ownership: it turns out that many investors have a 'home bias' and that many firms are still owned within a country (Stulz 2005). There is also the important question of *who* is financially globalized, *who* can participate in global finance or in institutionalized finance at all. There is the so-called 'digital gap': not everyone in every country has equal access to the new electronic technologies and digital media, and even if today more people have access to internet-based technologies than ever before, not everyone has the education and support that enable the more fortunate people to use the technologies as means of empowerment. For finance, this means that some participate in the electronic streams of money and information (for instance, people from Western countries) whereas others are excluded or cannot use the internet to empower themselves financially and economically. There is also what we might call a 'financial gap', including a 'banking gap'. Singh reminds us that 'half the world' is still 'unbanked' (Singh

2013). People who do not bank or who do not have access to banking either have too little money or save it in different ways. Many think this should be changed:

> Nearly half the people in the world aged fifteen and above do not have an account with a bank, credit union, cooperative, post office, or microfinance institution. ... The challenge for developing countries is to see how, with the help of technology and the effective design of money management services, this half of the population can be brought into the formal payment and banking systems. (Singh 2013, p. 42)

While it is not clear to me that currently existing and formal payment and banking systems are by definition to be seen as the ideal (this situation also opens up interesting and creative ways to 'do finance' in a different way; I will say more about this in the last chapter on alternative practices), this is an important qualification to the claim that finance is globalized. It may be globalized for some, but not for others. It still depends in which country, which region, and which part of the world you live. In sum, when it comes to finance and ICTs the territory is not abolished, and distance still matters; geography is not dead. It still matters for rights and well-being *where* things happen, *where* you do things, *where* you live.

This is also true for social space. ICTs bridge distance, but that does not mean that territory and proximity no longer play a role. Scholte argues that 'social space in today's world is *both* territorial *and* supraterritorial' because, for instance, people access the internet from a territorial location, aeroplanes need runways (one could say: *infrastructure* is still territorial or territorially anchored), 'global cities' still have 'a longitude and latitude' and global ecological changes have 'territorially specific impacts' (Scholte 2002, p. 26). If there is a 'global' and a 'local' sphere, they are at least very much interrelated:

> So social space should not be understood as an assemblage of discrete realms, but as an interrelation of spheres within a whole. Events and developments are not global *or* national *or* local *or* some other scale, but an intersection of global *and* other spatial qualities. The global is a dimension of social geography rather than a space in its own right. It is heuristically helpful to distinguish a global quality of contemporary social space, but we must not turn the global into a 'thing' that is separate from regional, national, local and household 'things'. (Scholte 2002, p. 27)

The same criticism is applicable to the relation between distance/proximity and communication. The local and the global should not be seen as mutually exclusive, separating the 'intimate' and 'immediate' from the 'distant'. Our binary thinking about 'global' *versus* 'local' relationships and spheres is misleading:

> Typically, local/global polarizations have depicted the local as immediate and intimate, whereas the global is allegedly distant and isolating. The local

> purportedly provides security and community, while the global houses danger and violence. The local is the arena for autonomy and empowerment, the global the realm of dependence and domination. The local is authentic, the global artificial. On such assumptions, numerous critics have rejected globalization with calls for localization. Yet these binaries do not bear up to closer scrutiny. After all, people can have very immediate and intimate relationships with each other via jet travel, telephone and Internet. (Scholte 2002, p. 28)

Thus, it turns out that social space is, and can be, both 'global' and 'local' at the same time. To stress this entanglement of the global and the local, sometimes the concept 'glocalization' is used (Robertson 1995). Originally a concept in Japanese business, Robertson introduced the term to the globalization discourse to refer to the blurring of boundaries between the local and global. For instance, even searching for 'home' and 'authenticity' are not *merely* 'local'; they are very much connected with globalization in the sense that they are a response to 'global' tendencies and can be interpreted as being part of the process of globalization. Globalization 'also involves the 'invention' of locality' (Robertson 1995, p. 35) and local culture gives meaning to global processes. Robertson writes: 'The global is not in and of itself counterposed to the local. Rather, what is often referred to as the local is essentially included within the global' (p. 35). The global and the local work together; there is one process of 'glocalization'.

A similar point can be made about the distance-proximity binary. Rosenau argues that in our era what seems remote seems close at hand at the same time, and that today world affairs are rather to be grasped as 'distant proximities' (Rosenau 2003, p. 3): an interaction between globalizing and localizing forces (p. 4). Note that he defines distance not only in objective terms but also and especially in subjective terms:

> Distance is not measured only in miles across land and sea; it can also involve less tangible spaces, more abstract conceptions in which distance is assessed across organizational hierarchies, event sequences, social strata, market relationships, migration patterns, and a host of other nonterritorial spaces. Thus to a large extent distant proximities are subjective appraisals – what people feel or think is remote, and what they think or feel is close-at-hand. (Rosenau 2003, p. 6)

What is 'close' or 'distant', then, is not so much a matter of 'objective' measurement but of subjective experience. For my discussion of financial technologies and distancing, this means that whether or not financial technologies create 'distance' or 'alienate' is not something that can be measured objectively; it depends on how people experience for instance global finance as mediated by ICTs – financial ICTs and others.

Let me now use these insights to address the question of how 'global' global finance is.

### 7.2.2. 'Global' finance? Lessons from social studies of finance

First, how purely 'global' are the global flows of finance? In the previous chapters I have used Castell's concept of the space of 'flows' to argue that global finance can be helpfully understood in terms of global flows of money and information. But Castells himself already remarked that most people (still) live in places: they perceive their space as *place*. There is the space of flows but also the space of places; both spaces co-exist. However, Castells sees a gap between these two forms of space: as noted in Chapter 4 he writes of two 'parallel universes whose times cannot meet because they are warped into different dimensions of a social hyperspace' (p. 428). Now the globalization literature reviewed in the previous chapter suggests otherwise: there are in fact many connections between the two spaces: the two spaces are entangled. The two spaces are not different dimensions – one 'free' immaterial and one material, place-bound – but rather seem to touch and mix in specific experiences and practices. They *do* meet in concrete ICT-mediated experience, including financial experience.

For instance, at first sight it seems that at financial exchanges a 'local' form of space has been replaced by a 'global' one: the local trading pit where traders were bodily present and communicated with one another verbally and non-verbally has been replaced by global electronic communication between traders everywhere in the world. It seems that a global world of numbers, a global sphere of information, has wiped out any place for the local – and indeed for human bodies. This seems to illustrate 'the death of geography'. Yet research in social studies of finance show a different picture of contemporary trading. It is true that screen-based technologies let things, people, and actions appear as numbers, whereas the trading pit involved 'full-body experiences' (Zaloom 2003, p. 263). But at the same time the traders are still located at a specific place and communicate with one another. There is an office, there are colleagues. As Knorr Cetina and Bruegger (2002a) have argued, traders have a double orientation: the global sphere is dominant – a sphere they access through the screen – but *at the same time* the traders also have a local orientation. The authors discern 'patterns of relatedness and coordination that are global in scope but microsocial in character and that assemble and link global domains' (p. 907). There are global dynamics, for sure, but also dynamics that have to do with two or more persons finding themselves in the same physical area where they see each other and are within each other's 'aural range' (Goffman quoted by Knorr Cetina and Bruegger 2002a, p. 908). Traders respond to other traders who are not physically present ('response presence'), which is made possible by electronic information technologies, but at the same time they have 'embodied presence' (p. 909) on the trading floor. There is a dematerialization of transactions, as Simmel also argued, but there is also a local presence and a local setting that is linked to embodiment and materiality.

This has implications for social space, for the relations between the traders. On the basis of ethnographical research on the trading floors of globally operating investment banks in Zurich, Knorr Cetina and Bruegger explore intersubjectivity in global trading. They distinguish between, on the one hand, the traditional face-to-face situation in which traders are able to observe one another since they are seated close together in the same physical space, and, on the other hand, what they call the 'face-to-screen' situation, in which traders face their screen (back-to-back rather than face-to-face). They describe how the latter spatial situation implies 'an orientation toward the screen that links the physically present person with a global sphere' but *also* 'a secondary orientation to the local setting and the physically present others participating in it', an orientation to the 'living presence of the trading floor' (p. 923). Traders maintain this 'double orientation' (p. 923) as they visually attend to the screen and at the same time hear other traders in the room (a situation made possible by material configurations – I will say more on this below). In terms of social space, these traders are thus 'global' and 'local' at the same time. There is an intersubjectivity that has to do with the mediated presence of other market participants on the screen and there is local interaction.

Moreover, embodiment plays a key role in these social spaces: *both* kinds of situations are bodily anchored. I will say more about embodiment in the next section, but it is already worth mentioning that the face-to-screen situation is less detached and disengaged than one may expect; trading in this situation and orientation also involves emotional engagement and even has an 'intimate' dimension. Cetina and Bruegger observe:

> Participants appear to be viscerally plugged into the screen reality and indicate this when they refer to market actions in terms of the penetration of their bodily preserves. ... Through their face and body front, traders reorient a significant fraction of their sensory equipment and bodily reaction capabilities to the life-form of the market .... Although traders are not able to slip through the screen and walk into this life-form, they stand within its intimate space, close enough to feel every tick of its movements and to tremble and shake whenever it trembles and shakes. (Cetina and Bruegger 2002a, p. 940)

As the authors remark, this physical response is not unlike what happens in the face-to-face situation. The traders not only have a double orientation, which we may call 'global' and 'local'; the 'global' orientation is itself also 'local' and 'intimate' in the sense of involving the bodily, physical engagement of the person. Both kinds of intersubjectivity are 'lived'. We could also say that what happens in the face-to-screen situation is in this sense as 'real' as what happens in the 'face-to-face' situation. The real/virtual binary blurs (see also the next section). It is a different kind of intersubjectivity and a different kind of sociality that is at play here, certainly, but when it comes to understanding these new forms of sociality and social space, the value of dualistic thinking is limited.

Second, to the extent that there are truly 'global' flows, these flows are still dependent on local places and infrastructures, and on the people who work at these places and with these infrastructures. This is also true for global financial flows. 'Global' financial practices are not located 'nowhere'; as Sassen has argued there are 'global cities' which form transnational networks and function as command centres for the global economy (Sassen 1991). As Knorr Cetina puts it, financial globality depends on 'bridgehead centres of institutional trading in the financial hubs of the three major time zones: in New York, London, Tokyo, and Zurich, Frankfurt or Singapore' (Knorr Cetina 2005b, p. 57).

Furthermore, as I already suggested in my reading of Cetina and Bruegger and as I will further discuss in the next section, 'global' finance depends on 'local' material infrastructure and arrangements. There are local computers and servers, and there are material arrangements in the trading room (e.g., making it possible for traders to sit back-to-back, side-to-side, face-to-face, etc.). Contemporary electronic trading depends on material wires as well, which connect financial centres situated at specific locations, and in high-frequency trading speed (and hence financial gain) depends on the location of your server. More generally, however 'virtual' 'cyberspace' may be, the internet and the World Wide Web and all electronic communications and transactions they make possible are entirely dependent on material infrastructures. Computers, servers, data centres, wires, etc., are located *somewhere*; they make possible 'global' communication and information, 'global' flows of money, but they are local and material structures. All this suggests that geography is far from dead when it comes to finance.

Note that this place-bound nature of global finance has implications for the vulnerability of the global financial system: to the extent that it depends on electronic ICTs, it is vulnerable in at least two ways: it depends on the functioning and security of its global electronic operations and it depends on the functioning and security of the *local infrastructures* that makes the electronic operations possible. However, since most citizens and indeed many people in finance are epistemically distanced from this more material and infrastructural side of finance, they may not be aware of these vulnerabilities.[1]

Note also that since the number of 'global' players in finance, for example in the field of HFT, is limited, 'global' finance is once again more 'localized'. Even if HFT firms are multinational corporations, it is good to realize that HFT is not happening only in 'cyberspace', somewhere 'out there', or 'nowhere'; instead, it is anchored in locally and territorially situated firms. These firms have physical addresses, material infrastructures, and they are located in specific countries and territories. There is not one homogeneous global financial space, but rather financial centres and hubs located in 'global' but at the same time 'local' cities. If there are global flows, they are not equally distributed over the globe. Some nodes in the network are bigger than others, and there are clusters.

Let me further explore these issues by moving to the field called 'geography of finance', which questions 'end of geography' arguments in relation to finance, and which instead highlights place (making) in finance. Combining philosophy of technology with more empirically oriented studies of the spatial dimension of

---

1 For more reflection on ICTs and vulnerability see my book *Human Being @ Risk* (Coeckelbergh 2013b).

money and financial technologies, geography of finance can help us attend to the moral and social significance and impact of financial technologies.

### *7.2.3. Towards a more refined geography of finance*

Some authors in geography of finance (and indeed in social studies of finance) define their view in opposition to Marx, Simmel, Weber and other writers, to whom they give the label 'modernist'. For instance, Gilbert argues that whereas modernist accounts such as those of Marx and Simmel take the view that money annihilates space, more contextualized understandings of money help us to see its geographical and dynamic dimensions. We need to situate money in time and space (Gilbert 2005). According to Gilbert, Marx argued that money played a role in the 'rationalization, homogenization and alienation of modern society' (Gilbert 2005, p. 362) but this 'fixation on money's role as a universal equivalent does little for understanding money as anything other than "the *nexus rerum* of capitalist society" ' (Fine and Lapavitsas quoted in Gilbert 2005, p. 362; see also Fine and Lapavitsas 2000, p. 366–67). Gilbert also criticizes Simmel's and Weber's view of money as an instrument in the rationalization and quantification of social life, and even Giddens' view of money as one of the disembedding mechanisms associated with modernity. Furthermore, 'modernist' accounts also assume that there is a natural, evolutionary development of the money form from primitive monies to national currencies. But, according to Gilbert, such accounts of money lack 'historical and geographical specificity' since they pay insufficient attention to 'the particular effects that money can have in different times and places' (p. 366) and 'the particular ways that spaces are ordered and places or territories are shaped through the circulation of money' (p. 371). They fail to address questions such as:

> How does money circulate? Who circulates money? Among whom does it circulate? What kind of meanings does money convey when it circulates? How do people use money? To address these questions requires case studies that take account of the social and cultural meanings in specific contexts and through distinct networks of social relations. (Gilbert 2005, p. 366)

This is a valid point. Yet in contrast to what Gilbert and others suggest, I do not think that these objections to Marx's and Simmel's 'modernist' view of money require *replacing* their accounts with empirical work in geography, social studies of finance, etc. 'Modernist' or not, Simmel, Marx, Weber, and other critics of modernity provide valuable insights and powerful interpretative and normative frameworks without which any work on financial technologies remains uninspired, directionless, shallow and incomplete. Moreover, Gilbert sketches a caricature of the criticisms of modernity which are much richer and nuanced than she suggests. That being said, addressing the main question of this book concerning the relation between financial technologies and distancing requires paying attention to the spatial and historical dimensions of these technologies, and to their use and meaning in specific contexts. This means that empirical and empirically oriented studies of the geography of finance and financial technologies are a helpful and even necessary complement to philosophical and theoretical analysis. (And more generally, philosophers of technology can learn from studies in other disciplines that teach us more about the relation between specific technologies and space.) In the previous sections I already used work from more empirically oriented disciplines and fields such as globalization studies and social studies of finance. Let me now continue my inquiry into how space has been (and is being) changed by new financial technologies, with the aim of arriving at a more refined geography of financial technologies.

How then does contemporary finance and its technologies *change space*? Some authors in geography of finance stress the global, 'flow' aspect of finance. Pryke and Allen argue that new forms of money lead to what they call 'a new monetization of time-space' (Pryke and Allen 2000, p. 264). What does this mean? In contrast to for instance Gilbert, they still see a role for Simmel's theory. They also pay attention to specific financial instruments and their relation to *technology*. They want to know what new financial instruments such as derivatives mean for our experience of space and time and for the rationalization of money and risk, and argue that technology plays an important role in these new developments:

> The use of new technology ... has not simply enabled the cross-border flow of information and use of databases, but has allowed the complicated pricing of new instruments and the calculation of 'cross-border arbitrage that integrates financial markets around the world'. (Pryke & Allen 2000, p. 268)

Although financial risk is not new, what has changed is the ways risks interact and are transmitted through 'electronic highways', thus changing social space (p. 269). The authors argue that informatics in particular have impact on how we conceptualize time and its relation to space and, ultimately, on the form of our life. In evaluating this impact, they agree with Simmel that the pace of life is increased and that today's money culture also renders space quantitative and calculable (p. 270). Derivatives have influence on this 'rhythm of money' (p. 271) and society. They are what Simmel would call a 'sociological phenomenon'. The technology has changed, and therefore the timing and spacing of our daily lives have changed:

> Simmel's 'streams of money' have become technologically empowered torrents of 'digital capitalism'. This new stream mixes domestic and international flows in a way unheard of in Simmel's time. The everyday – from work, to consumption, leisure, pensions and so on – is now a criss-cross of spaces and times which alter the pace and thus experience of life in late modernity. ... The numerous forms of money, linked through new technologies, are arguably responsible for re-coordinating the times and spaces of the everyday in 'real' consequential ways .... (Pryke & Allen 2000, p. 275)

This means, according to the authors, that we now perceive a different world: 'the combined effect of the intensitivity of information and the growing reliance upon electronics as a way of seeing the world (particularly in financial centres)' gives us a new imaginary, one 'fuelled by electronics and the terminal' (p. 278), one of speed and one of flows.

But does this mean that this new world of flows is situated 'nowhere'? Do these new financial instruments and financial technologies bring about 'the end of geography'? Interestingly, the authors argue that the new imaginary they describe does not annihilate space but instead potentially recomposes it: it 'holds the potential to recompose, *re-rhythm* 'real' geographical spaces as financial calculations unwind in the everyday, far away from the terminals in financial centres'. (p. 279). Thus, space is not destroyed but changed, re-ordered. Moreover, the emphasis of the authors is rightly on the relation between what happens in the financial centres and the everyday. Their point is that today our perception and use of money has changed. Global finance is not something 'out there', but shapes everyday action and experience. Particularly interesting for my reflection on 'distance' is the observation that what happens in the space of 'flows', the space of 'calculations' and complex financial instruments, is not merely something 'technical' or something that belongs to the world of 'finance' as if 'finance' is a box separated from 'society' and 'life'. Like Simmel, Pryke & Allen suggest that the lifeworld is deeply and importantly shaped by finance and financial technologies. The flows on the 'electronic highways' of finance change the spaces and rhythms of our daily lives.

However, in Pryke & Allen the more concrete and 'local' face of these changes to daily lives remains somewhat vague. And within geography of finance there are also objections to the imaginary of a global space of flows, or, at the very least, accounts that complement the 'global' image with a 'local' one. Many other geographers think that today money has *not* been deterritorialized or at least (still) has an important local and territorial aspect. In an issue of *The Industrial Geographer* on the geography of finance, for instance, the editor remarks that it is a 'folly' to assume that 'capital and information flow frictionlessly to the locations which promise the greatest return' (Graves 2004, p. 1). Others have pointed out that the 2008 financial crisis had not only its 'global' but also its 'local' geographies (Martin 2011). This is especially so in terms of the social and economic impact of what happens in 'global' finance. Martin argues that while it was true that something that was local in origin – the housing bubble in the US – had global consequences, the 'global' crisis had locally varying impacts. He calls the crisis therefore 'a stark example of "glocalisation" ' (p. 589). On the one hand, financial space has been globalized in the sense that local financial outcomes depend on what goes on at far removed locations and relations. On the other hand, global financial transactions and markets (still) depend on local financial conditions. Martin shows that in the UK, for instance, the London economy suffered less from the crisis than expected; instead, the West Midlands (a manufacturing, not a financial region of the country) was hit hardest.

With regard to responsibility, it is important to make such relations and impacts visible. Unless they become visible, ascribing and exercising responsibility is difficult. Both citizens and professionals in finance need to be more aware of the concrete, place-bound, 'local' impact of their allegedly 'global', place-less actions in the world of 'flows'.

Another example of how global finance has local impact is speculation with food. A Foodwatch report called *The Hunger-Makers* (2011) argues that speculation with food on commodity exchanges drives up prices at the expense of the poorest: 'Speculation on commodity exchanges with food products such as corn, soybeans and wheat is strongly suspected to contribute to poverty and hunger' (Foodwatch 2011, p. 6). The report argues that since ancient times speculation with grain has always been morally condemned (and rightly so), and that today action should be taken to protect humans from hunger:

> Financial managers at exchanges and investment banks, who maximize their sales volumes and fee revenues with the help of commodity markets, thereby potentially causing humans to suffer from hunger, perhaps even die, should prove that their business activity does no harm. (Foodwatch 2011, p. 65)

Again financial activity in one place has potential effects – 'local' effects – in other places. This is about actions at financial centres – e.g., futures exchanges in London, New York, Chicago, Paris, Dalian, Shanghai, and Mumbai – which have impact on impoverished people who spend 80 per cent of their income on food, people located in Asia, Africa (e.g., Uganda and Burkina Faso) and Central America. For instance, poor people in Central America saw their tortillas become 70% more expensive within a year (Foodwatch 2011, p. 12). This speculation was made possible by ICTs: the massive and globalized use of financial products such as futures is only possible with advanced technologies such as electronic information and communication systems.

Again these spatial aspects and effects of financial technologies have implications for responsibility. To the extent that place still matters in 'global' finance, then with regard to responsibility in finance it is important that both citizens and people working in the financial sector become aware of financial-geographical relations such as those highlighted by this Foodwatch report. As said in Chapter 6, one of the conditions for ascribing and exercising moral responsibility is knowledge. Currently, the technologies used in the financial sector show what 'the market' does and enable traders to tap into the world of flows; but they hide the concrete, place-bound consequences for people elsewhere in the world who may suffer from price changes in food markets, oil markets, etc.

These are just a few illustrations of how reflections on financial technologies, space (e.g., distancing), and ethics of finance may benefit from more empirically oriented work in fields outside philosophy. And important for the aim of this chapter: they also show how the new world made possible by financial technologies is both global and local, and in any case less homogeneously global than I suggested in Chapter 4. Reflection on the social and moral consequences of new financial technologies need to take this geography into account.

### 7.3. How 'immaterial', 'disembodied', and 'virtual' is global finance? The material world of finance

Let me now say more about the alleged 'immaterial', 'disembodied', and 'virtual' dimensions of global finance. The previous chapters gave the impression that global finance is 'immaterial' and 'virtual', as a result of electronic technologies and digital media which seem to create a separate, virtual sphere that is not only disembedded (disconnected from the local) but also disconnected from the material and the body, and remote from the real world. But there are good arguments to qualify this claim. Let me draw on social science and philosophy to develop this point.

#### *7.3.1. Anthropology and STS: Networks of humans and non-humans*

In the previous section I already said that global finance depends on material infrastructure: networks, cables, satellite connections, trading floors, terminals, screens. In this section I show that the role of these artefacts (and built environments) in global finance can be better understood by using the conceptual apparatus of anthropology, science and technology studies (STS), and social studies of finance. The work of Callon and Latour is especially influential in this area. They use the term 'actor-networks'. Actor-network theory (ANT) analyses technological objects as part of the social. 'Technology' is not separate from the social; instead there are networks of humans (actors) and things ('actants'). This concept has been used to describe scientific practices. STS has shown that what scientists do in their laboratories can be better understood by using this concept and approach. But similar work has also been done in the domain of economy and finance. During the past 15 years, anthropologists and sociologists of science and technology have increasingly studied economic and financial knowledge and practices. Turning from the laboratories of the scientists to the 'quasi-laboratories' (Callon 1998) of economic and financial agents, authors such as Callon, MacKenzie, Knorr Cetina, and Zelizer have demonstrated that economy and finance can be understood as technical practices, as practices mediated by technologies. Markets have their *devices* (Callon 2007).

This approach helps us to highlight and understand the material dimension of new financial ICTs and the way they change financial practices. At first sight, the new ICTs seem to be all about immaterial 'information'. It is therefore tempting to conceptualize their nature and influence in terms of 'information'. For instance, Floridi works on the philosophy and ethics of information (Floridi 2013) and one could apply this framework to financial information technologies. However, next to other problems discussed in Chapter 5, this approach invites the objection that there is a significant disadvantage to using 'information' as a central concept if it means that little attention is paid to material artefacts and their relation to people. The very concept of 'information' hides the material side of the new technologies. Pinch and Swedberg write:

> Though information may seem 'non-material', in reality this type of technology permits new forms of entanglements between people and objects and can crucially change the material circumstances whereby exchange of goods and knowledge occurs and when things and ideas circulate. (Pinch and Swedberg 2008, p. 12)

Thus, if the new technologies change financial practices and, ultimately, the entire economy and society (they become 'informational') then this does not mean they become 'immaterial'. Instead, the new technologies and the 'information society' entail *different* human-material entanglements. Furthermore, by calling attention to the material-technological dimension of finance, these authors question not only the apparent immateriality and virtuality of finance, but also the abstraction of the classical economic agent as found in economic theory: the *homo economicus*. In *The Laws of the Markets* Callon writes: 'Yes, homo economicus really does exist. ... He is formatted, framed and equipped with prostheses which help him in his calculations and which are, for the most part, produced by economics'. (Callon 1998, p. 51) Economics thus 'produces' the agents and abstractions it assumes. In Callon's approach 'prostheses' and 'devices' include not only material-technological artefacts but also economic theories, methods, and techniques that assist calculation and indeed produce the calculative culture we live in. Or rather, to put it more precisely and geographically, economics produces calculative spaces. (There is still an inside and outside, and we may be able to change these spaces; I will return to the topic of social-technological change.)

So what is the structure of the relations between people and artefacts, between humans and non-humans? Which concepts can we use to describe financial practices? As in science studies, the *network* metaphor has been used to describe financial-social systems and practices. For instance, Leyshon and Thrift (1997) have identified actor-networks in global finance such as nation states and machine intelligence, which order people, texts, money, and other non-human elements (p. 298). This enables us to paint a different geography, a more local and material one. Even if we define money in terms of information, it is important to look at the social, cultural, and geographical embeddedness of the 'flows': money as information is then 'circulating in specific, separate but overlapping actor-networks, made up of actors, texts and machines, which think and practise money in separate but overlapping ways'. (p. x) What does this mean?

For my discussion of financial technologies it is interesting to first focus on the claim about technological mediation, on the *machines*. Barry and Slater write:

> economic transactions increasingly take place through technological mediation, and these technologies are not neutral tools of economic policy and practice. The contemporary prestige of such terms as 'network society' and 'information economy' point to the role of technologies in conceptualising and reconfiguring economic action. (Barry and Slater 2005, p. 7)

With this point about the non-neutrality of financial technologies, the authors find themselves in agreement with contemporary philosophy of technology and indeed with the approach of this book: financial technologies are not neutral instruments that are only of 'technical' interest; instead, they co-shape our moral and social world. However, I take issue with Barry and Slater's suggestion that technological mediation is mainly a contemporary phenomenon. As I have argued in Chapters 2 and 3, financial technologies have always mediated financial practices and have helped to shape society. Of course *the kind of technologies* change over time: today, electronic ICTs play an important role in finance, in the economy, and in society.

For thinking about ethics and responsibility in finance, then, it is important to realize that not only humans but also artefacts-as-used-by-humans (and sometimes, artefacts that take over from humans) play a role in creating the conditions under which responsibility problems arise and under which responsibility is ascribed and exercised.

Another important and related point is the materiality of financial technologies. Since the STS approach highlights specific technological mediations, we see that even global finance, which seemed so virtual and a matter of abstract 'streams', is at the same time dependent on very concrete, material artefacts, processes, networks, performances, and infrastructures. Let us further look at this through the lens of social studies of finance.

### *7.3.2. Social studies of finance: Trading floors, cockpits, and mining*

Knorr Cetina's study of trading floors at foreign exchange markets in Zurich and New York is a good illustration of how a more 'anthropological' method enables us to get insights into the material aspects of finance. This is an example of a description of trading floors with their technologies:

> In doing deals, all traders on the floors have a range of technology at their disposal; most conspicuously, up to five computer screens, which display the market and serve to conduct trading. When traders arrive in the morning they strap themselves to their seats, figuratively speaking, they bring up their screens, and from then on their eyes will be glued to these screens, their visual regard captured by it even when they talk or shout to each other, their bodies and the screen world melting together in what appears to be a total immersion in the action in which they are taking part. (Knorr Cetina 2005a, p.p. 124–125)

Yet this description does not only show that ICT plays an important role in these practices (these are electronic markets); it also shows that this is a material practice and that humans and technological devices 'merge' in this financial practice. As in many other computer-mediated practices, humans and computers 'collaborate' or even 'merge', effectively becoming like 'cyborgs': blends of humans and machines. Moreover, 'the market' is not entirely abstract and immaterial; it gains presence as numbers show up on the screens. What is called 'the market' is constructed by these terminals, dealing and information systems, electronic accounting services, etc. It also includes the trading floors themselves. Using terms such as 'the market' and 'information flows' hides this material and technological dimension. This is problematic since the technologies do more than function as instruments. What they also do is what Knorr Cetina calls 'delivering a world' (p. 126) or even a 'life form' (p. 127). Technologies shape our thinking and they also create new realities. The computers and screens are more than tools or media, they contribute to the creation of 'the market' and thus – to speak with McLuhan – they have their own 'message':

> Traders are not simply confronted with a medium of communication through which bilateral transactions are conducted, the sort of thing the telephone stands for. They are confronted with a market that has become a 'life form' in its own right, a 'greater being', as one of our respondents, a proprietary trader in Zurich, put it. (Knorr-Cetina 2005a, p. 127)

This approach gives us a more diverse phenomenology of finance than the 'flows' or 'networks' of Castells or 'network markets': it does not necessarily discount 'flows' and 'network' experiences, but in any case *also* reveals how *material* technologies contribute to shaping the world of finance:

> From the traders' perspective, and from the perspective of the observer of traders' lifeworld, the dominant element in the installation of trading floors in globally interconnected financial institutions is not the electronic infrastructural connections ... but the computer screens .... The market on screen takes on a presence and profile in its own right ... It is not simply a 'medium' for the transmission of pre-reflexive interactions. (p. 129)

Thus, on the one hand there is the experience and presence of the market on the screen, which takes on 'a presence in its own right'; on the other hand, there are the technologies and indeed material artefacts and infrastructures that make possible this experience. Traders and their 'markets' remain dependent on the material infrastructure, the 'pipes'. It seems that markets and, more generally, financial practices have both a 'metaphysical' and a physical, material aspect. As Knorr Cetina acknowledges, the market-screen has a 'flow' character. Information flows, but the market is also flowing. The trader's lifeworld is also in flux. But this flux becomes present, or perhaps comes into being, through the technical devices.

The market-in-flux is present on the screens of the traders. In that sense, financial 'flux' and 'flow' is both spatial and non-spatial, material and non-material. The geography and 'message' of finance is in that hybrid, multifaceted medial reality.

Moreover, social studies of finance also reveal global finance as embodied, even if it deals with numbers. Zaloom has argued that there is a physicality to market numbers (Zaloom 2003). Finance is a material practice but also a physical, bodily practice. This was of course especially the case in the trading pit: 'for pit traders, both delivering and receiving the bids and offers of the pits are full-body experiences', which means that 'the pit requires stamina and strength'. The traders' voice and stature are important, and their physical location matters. For instance, it is important to be able to attract the attention of others, and this is difficult for anyone who is physically short and slight. One has to create a presence in the pit (Zaloom 2003, p. 263). Screen-based trading, by contrast, seems a lot less 'embodied'. What is 'screen-based' trading? Drawing on fieldwork on futures markets in Chicago and London, Zaloom observes that new electronic trading technologies have supplanted traditional methods where traders meet; traders are now using screen-based technologies and futures traders act with 'the space of flows' (p. 258). However, even in this kind of trading there are also material things involved and screen-based trading is still embodied. To quote Knorr Cetina and Bruegger (2002a) again: enabled by the electronic technologies traders respond to other traders who are not physically present but at the same time they have 'embodied presence' (p. 909). In one trader different kinds of orientations and presences meet.

Another concept used to highlight and describe the materiality of financial practices is the notion of 'assemblages'. Muniesa et al. proposed the notion of 'market device' to refer to 'the material and discursive assemblages that intervene in the construction of markets' (Muniesa et al. 2007, p. 2). They use the term 'agencement' to emphasize the distribution of agency and materiality. For instance, if there is 'calculation' in finance – as for instance Simmel also argued – then these authors argue that it is 'the concrete result of social and technical arrangements': market devices 'affect the ways in which persons and things are translated into calculative and calculable beings' (p. 5). For instance, derivatives make possible increasingly complex markets.

Using this approach, we can describe and interpret financial practices as they are shaped by electronic technologies. Beunza and Stark (2004) have studied arbitrage in a Wall Street trading room in order to analyse 'trading in the era of quantitative finance' (p. 369). On the one hand, the kind of technologies and sciences used in these kinds of financial practices suggest again the 'flow' image and the idea of a 'virtual' world. The authors write that in electronic markets, where 'powerful computational engines' (p. 370) and mathematics play an important role, actors are 'immersed in a virtual flood of information' (p. 369). On the other hand, this does not mean that that there is no 'human' work involved (see also below); humans select and interpret. Moreover, at the same time it is also a social, material, and 'local' practice. Let me say more about this particular study in order to elaborate on the role of recent financial technologies.

Beunza and Stark say that they have conducted ethnographical field research in order 'to explore the socio-cognitive, socio-technical practices of arbitrage' (p. 373). The trading room is revealed as 'an assemblage of instrumentation' (p. 374). This assemblage includes computer programs which execute automated trades: 'robots' (Beunza and Stark) or 'algorithms' (the term used today). Studying the trading room as a 'laboratory, the authors observe how important instruments are for the trade. Traders seek opportunities, and instruments visualize these opportunities, and calculation is delegated to robots. Beunza and Stark argue that cognition is distributed among networks of tools, including 'computer programs, screens, dials, robots, telephones, mirrors, cable connections, etc.' (p. 389). But the instruments are always more than tools: they also shape how we perceive the world and how we act. Inspired by Latour, Beunza and Stark write:

> we found our trader's tools remarkably close to Latour's (1987) definition of scientific instruments as inscription devices that shape a view. Scientific instruments ... display phenomena that are often not visible to the naked eye. ... Similarly, the trader's tools reveal opportunities that are not immediately apparent. (Beunza and Stark 2004, p. 390)

There are various visualization techniques. Like the laboratory, the trading room gathers instruments and techniques. The trader's workstations (e.g., from Bloomberg) – both the physical screens/terminals and the software – visualize financial data and provide information on trade execution. But when they do this, they are more 'active' than one would expect and shape the trader's world: 'Screen instruments are not mere transporters of data, but select, modify and present data in ways that shape what the trader sees' (p. 390). Moreover, the computer shares not only in the cognition but also in the agency of trading. The computer calculates when to trade and shows numbers and colours on the screen for that purpose. This kind of trade thus becomes increasingly 'robotic'. Beunza and Stark write that financial robots are first programmed: human knowledge becomes algorithms and computer code. Then the robot is 'piloted', taking into account a kind of 'traffic control – cues and signals from other parts of the room' (p. 393). Indeed, the authors compare piloting the robot with engaging in car traffic. Traders engaged in 'automatic trading' and 'high-speed trading' work in what we could call a 'trade cockpit':

> Automatic trading poses the same challenge as driving a car at a high speed: any mistake can lead to disaster very quickly. ... As with Formula 1 car racing or high-speed power-boats, traders need excellent instrumentation. Indeed, they have navigation instruments as complex as an airline cockpit. (Beunza and Stark 2004, p. 393)

Again, we might want to use the imagery of the 'cyborg' to describe this configuration: both the car race pilot (or jet fighter pilot, race boat pilot, etc.) and the pilot who is taking part in financial races are entangled with their electronic technologies and completely depend on it for their activities. But the actions are not taking place in social isolation. This piloting, with the help of instruments, is then combined with social interactions. Screens provide the trader 'with as many dials as a cockpit in an airplane' (p. 394), but traders also 'interpret cues from nearby desks to gauge when to take a particular security out of automated trading' and thus 'remain entangled in the social relations'. In this sense, the tools are 'socio-technical' (p. 395).[2]

Thus, these and other 'social studies of finance' work convincingly to show that global finance is less homogeneously 'virtual' and 'flow'-like than it seems; it has also a material, bodily, and social side to it. Seen and experienced from one perspective, quantitative and electronic finance is global and abstract, but analysis of 'the entanglements of actors and instruments in the socio-technology of the trading room laboratory', studies of the 'tools of the trade' (p. 397), show that seen and experienced from another perspective, the practices are also at the same time human, material, and local. (In the next section I will say more about that 'human' aspect.) Keeping in mind the topic of the previous section, we can also observe that this approach to finance enables us across the global-local distinction. As for instance the work of Knorr Cetina shows, it is possible to understand global finance by analysing the microsocial structures of financial practices (Knorr Cetina 2005a and 2005b; see also Barry and Slater 2005, p. 11). The 'local' screen, for instance, makes the 'global' market visible and present (Knorr Cetina and Bruegger 2002a).

Using this approach, we can now re-interpret some of the financial technologies introduced and discussed in the previous chapters, for instance (1) the algorithms used for high-frequency trading and (2) Bitcoin. The former has already been discussed by relying on Beunza and Stark: it turned out that the abstract instruments of computational finance were linked to very material tools such as screens and other parts of the trader's cockpit which together made possible combined (or perhaps even merged, 'hybrid', 'cyborg'-type) human-machine cognition and action, and which shaped the trader's perception of what to do. But Bitcoin still needs our attention. How 'material' is Bitcoin?

Bitcoin involves an 'algorithm', which seems to be non-material: it appears as 'code'. But again we can look at how this 'immaterial' face of the technology is made possible by an assemblage of material artefacts and infrastructures (and humans working with them). These assemblages are often obscured, and this is also the case with Bitcoin: bitcoins appear 'digital' or even 'virtual' to us, but there is material equipment such as computers and servers that make possible their creation and processing. Its 'flows' are dependent on these material artefacts and

2 The authors also remark that this social aspect includes social-spatial organization: distance is used to organize diversity, but there is also proximity to make sure that cues when to turn off the robots can be provided.

infrastructures. Furthermore, Bitcoin is conceptualized as something immaterial, but its 'ideology' and the way it is technically set up has a 'materialist' side to it. Maurer et al. discuss what they call the 'practical materialism' of Bitcoin (Maurer et al. 2013, p. 262). They argue that there is materialism in the way Bitcoin is conceptualized and how bitcoins are mined: 'Bitcoin brings matter back into the picture, and vitally so: in the code and in the computer "rigs" with which coins are "mined".' (262). What does this mean? The authors argue that Bitcoin materializes values such as privacy and liberty (through cryptography and decentralization), and that it refers to metal with its metaphor of 'digging' and analogies to gold. As in the case of metal money such as gold, there is also a limited supply and thus inflation is avoided. This limited supply is materialized in the way Bitcoin works. Like metal digging, making bitcoins requires expending of resources. It requires electricity and infrastructures (e.g., the electrical grid, the internet), and human and computer labour (p. 271–272). The authors call this Bitcoin's 'digital metallism' (p. 269): it is digital currency, but its process resembles that of metal money. Perhaps we could say that Bitcoin 'comments' on metal money such as gold, indeed 're-mediates' it (see Chapter 3). As 'digital' money it does new things (e.g., the peer-to-peer, decentralized aspect), but at the same time it also wants to be able to do what gold did and does when it comes to trust. By limiting supply it imitates the kind of trust in metal that led to the gold standard. Indeed, Bitcoin is and remains social in so far as it impacts on, and projects a view on, 'the social dynamics of community and trust' (p. 262). Users are supposed to trust the Bitcoin algorithm itself, instead of governments and corporations. Bitcoin also 'promises' socially relevant values. Maurer et al. write: 'The point is not whether Bitcoin "works" as a currency, but what it promises: solidity, materiality, stability, anonymity, and strangely, community' (p. 263). I will say more about these promises in the next chapter. Let me first call into question the image of contemporary finance as non-human (next section) and impersonal (last section).

## 7.4. How non-human, machine-like, and calculative is global finance?

Is global finance only about abstract numbers and calculation? Is there no place for the human in what seems a world of algorithms? Are we victims of 'the system', imprisoned in the realm of calculative reason? Is our society determined by the cage that we have created ourselves with our technologies? While Simmel and Weber were right to point to the dangers of calculative rationality and bureaucracy, global finance is not quite the 'iron cage' (Weber 1905, p. 123) it might seem for at least the following reasons.

First, of course there is calculation, but it is not 'mechanical' calculation; as Beunza and Stark have argued (2004, p. 371), there is 'judgment'. Traders interpret (Zaloom 2003) and judge, they are not mere cogs in the machine. Having studied electronic futures markets, Zaloom argues that on the one hand finance has always been based on numbers. As I also suggested in the first chapter, numbers have been part of the history of financial technologies: 'From the invention of number-based accounting practices such as double-entry bookkeeping, numbers have been a cornerstone of economic calculation, providing the essential tools for rationalized action' (Zaloom 2003, p. 259). On the other hand, when we examine the knowledge practices of traders and attend to the *consumption* of numbers, 'calculation' is not the best and only way to describe what is going on. Zaloom argues that 'traders' practices are best described as interpretation rather than exacting calculation' (p. 259). In particular, Zaloom argues that traders 'undermine the rationalizing effects of technology' by bringing out the social dimension of the market:

> [Traders] interpret the market numbers through the particular framing of each technology and thereby unearth the specific social dimensions of market conditions. Traders bring questions about the social content of the market to their calculations no matter how much software designers try to remove such cues from their programs. Who are the competitors? What are their individual styles? Are they scared, stolid, eager, or anxious? ... Social contextualization and interpretation are critical parts of traders' calculations. (Zaloom 2003, p. 261)

This does not mean that the technology makes no difference; on the contrary, the social in face-to-face situations is different from the social in face-to-screen situations. Whereas in trading pits the social epistemology is local, screen-based trading makes possible a different form of sociality. But I interpret Zaloom's descriptions as showing that traders use their imagination to re-humanize and *re-personalize* (and indeed re-socialize) the market and the abstract numbers:

> On the screen, traders imagine and identify competitors within the changing numbers. They construct a digital landscape of social information. These competitors are cloaked in the abstracted numbers of the market, but traders assign personalities and motivations to the characters behind them. ... Software designers may attempt to excise social information from their technologies. But traders create new social contexts to replace the ones they have lost. (Zaloom 2003, p. 261)

Thus, the electronic technologies *hide* the social, the human, and the personal dimension but traders re-socialize, re-humanize, and re-personalize the numbers: by means of interpretation and, as Zaloom also writes, by talking to coworkers (p. 261). Note, however, that although there are these personalization and socialization processes, the electronic technologies themselves still play an important epistemic role in hiding the social. Moreover, although the personalization and socialization processes bring back the social and personal in the form of the *competitors* and their (imagined) intentions, this is a particular framing and selection of the social. Traders are persons who are more than 'competitors' on the market, and at the end of the day there are many *other* people who remain out of sight: the stakeholders, the people who invest, produce, and work in the businesses related to the trade, but also the people whose lives are influenced by the price changes due to the trade, such as changes in housing prices and changes in food prices. This was already the case in the face-to-face situation of the trade pit, of course, but now the electronic technologies promote a further globalization and thus a further distancing.

Second, more generally there is meaning-giving in contemporary financial practices and this is a further sense in which they are or remain 'human'. Allen and Pryke argue that the 'flows' of contemporary finance are always *lived*, *interpreted*, and *experienced*:

> rather than talk of a single objectified culture of money, it is perhaps best to talk about its flows and movements as something which is lived, experienced, and interpreted by particular groups in particular places at particular times. On this view, money cultures are made up of people who position themselves in relation to the circulation of money and are also positioned by it. (Allen and Pryke 1999, p. 65)

This opens up the possibility of having other experiences and attitudes towards finance, and perhaps new forms of sociality (see the next chapter). There is no single trading experience, no single financial experience, and it is also important to pay attention to 'what people *imagine* themselves to be involved in when they experience the gyrations of advanced forms of digitalized money' (p. 66). Allen and Pryke conclude that there are distancing effects and there is a change in rhythm and circulation, in the sense that what happens in the financial world is analogous to what happens in the everyday world of circulation, but that in the end it is people (that is, *humans* and persons) who invest powers in these flows and they can position themselves in relation to them. Thus, there is neither determinism nor autonomous development of the technologies and the technological practices under considerations.

For ethics and responsibility, this human and subjective aspect implies that there is room for variation and change. Since people play such an important role in global finance, what matters in terms of responsibility is also the *culture* of specific practices which, in principle, could be changed. For instance, there are specific *risk cultures*. STS and anthropology can help to study these cultures and reveal more variety than presupposed by Simmel, Weber, and other critics of modern society. Such cultures, however, remain part of modernity and should *also* be studied as such. For instance, Green (2000) has explored the modern notion of risk within global financial markets. He argues that modern risk (management) is still part of 'the still confident modern project of social attempts to control nature and the future' (Green 2000, p. 78), an ambition which he thinks is shared by financial markets (p. 81). Financial markets commodify risk: they put a price on it. At the same time they also create risk and uncertainty, which needs to be managed. Cultural practices develop around this risk production and risk management. We can study these particular practices using all kinds of concepts, but it is important to keep in mind that they are part of our *modern* way of dealing with money, risk, and people. This means that the Simmelian framework constructed earlier in this book remains relevant.

Third, STS also shows that calculation is not an abstract process divorced from people and things. In *The Laws of the Markets* (1998) Callon does not write about quantification and calculativeness in general and in the abstract, as Simmel did, but asks the question 'Under what conditions do calculative agents emerge?' (Callon 1998, p. 4). This approach has led him to emphasize and study the social and material aspect of calculation:

> Calculating ... is a complex collective practice which involves far more than the capacities granted to agents by epistemologists and certain economists. The material reality of calculation, involving figures, writing mediums and inscriptions ... are decisive in performing calculations. ... calculations are made in the *quasi-laboratories of calculative agencies* .... (Callon 1998, pp. 4–5, Callon's emphasis)

Thus calculation is itself a social-cultural process (rather than 'has a social-cultural context'), is a collective process, and is mediated by various technologies. Moreover, as I said previously, like natural sciences the science of economics also shapes what it describes. Interpretation and theorizing is not a detached, neutral 'academic' exercise; it has effects on the worlds and on agents. Callon suggests that the *homo economicus* is also *made* by economics. And the market is not something external to the social or to the human, or something that belongs to a separate technological or economic domain; it is something human as well. Humans study the market (and thereby shape it) and humans act to bring it about: 'The market is no longer that cold, implacable and impersonal monster which imposes its laws and procedures .... It is a many-sided, diversified, evolving device which the social sciences as well as the actors themselves contribute to configure' (Callon 1998, p. 51).

This is also true for the *homo economicus*: if this type of human is *made* by humans, then humans can also act differently. Influenced by Callon, MacKenzie and Millo (2003) have questioned the view that in finance we can (only) find the *homo economicus*. In their study of options trading at the Chicago Board Options Exchange, they argue that the people they have observed did not act as 'atomistic, amoral *homines œconomici*' (MacKenzie & Millo 2003, p. 109). Instead, reputation and interpersonal trust were important. The authors argue that 'material means of calculation' (p. 140) play an important role in economic action, but so does 'regard'. They even suggest that even autonomous software agents, while amoral, might not simulate the cognitive capacities of the *homo economicus* and have a more open construction (p. 141). The 'human' nature of finance and of its technologies thus opens up the possibility of social-financial change.

If this is right, then the calculativeness of our culture, which Simmel rightly criticized, is neither *determined* by technology nor only a matter of calculative *thinking* alone. 'Technology' or 'economy' or 'finance' is not something external to human action and human culture. If we want to change finance, then our task is not to change a kind of monster that is external to the sphere of human and material-technological action. If we want to change finance, we have to change humans and we have to change technologies. We have to change both people and things. We have to change entire networks of humans, technologies, and knowledge, which have to be generated and re-generated over and over again. Again, this frees up space for social-financial change. If finance has to be *performed*, then it can be performed differently. Moreover, the technical dimension of finance is not only a matter of *machines*; many more types of artefacts and media are involved, and changing global finance means intervening in the full range of these assemblages.

If there is alienation, then this is not only a broad cultural-structural process, or a development made possible by just one specific technology; it is also something that takes shape in these performative processes which are both human and non-human, and which are dependent on a variety of technologies, media, and infrastructures. The various forms of distancing identified in the previous chapters do not constitute an 'iron cage' which is forced on us from 'above' or from 'outside'. These social and anthropological studies show that the moral and social geography of finance is far more 'horizontal' and endogenous to the socio-technical world than assumed by the idea of 'the machine' (here: the financial machine, the 'money machine') which alienates us. That being said, this does not mean that seen from this angle there is no longer distancing or alienation; instead, the processes of distancing are just much more part of what we do – in finance and elsewhere – than may have been suggested by the Simmelian distancing thesis constructed earlier in this book. In global finance there is abstract calculation, and there are numbers and abstractions, but these abstractions are at the same time part of *lived* financial culture. In spatial terms, there is distancing, but this distancing is shaped in concrete social-material practices.

To conclude this section: both STS and Simmel help us to recognize the important role of financial technologies and media in shaping our society and culture. But in contrast to Simmel, STS and related approaches help us to look in more detail at the material aspect of financial technologies (instead of only considering for instance their distancing and 'calculative' aspects in the abstract) and to attend to the various ways in which humans and financial technologies are entangled and *bring about* calculativeness. Moreover, there is no homogeneous calculative culture or society, but calculative cultures and spaces that do not necessarily cage the entire culture or lifeworld.

## 7.5. How impersonal is global finance, and how personal can it be?

Anthropologists have also contested Simmel's claim about the impersonal character of social relations due to money. Against Simmel, Hart (2007) argues that money is always personal and impersonal. While he acknowledges that money was traditionally impersonal, he argues that there have always been more personal forms of money:

> Money was traditionally impersonal so that it could retain its value when it moved between people who might not even know each other. ... Money in this form is an instrument detached from the person who uses it. ... Bank credit has always been more directly personal, being linked to the trustworthiness of individuals. (Hart 2007, p. 12)

Moreover, there are forces of de-personalization but also forces of what I propose to call *re-personalization.* Hart does not deny that there are forces that try to de-personalize the economy, including for instance labour. In capitalist societies money stands for 'alienation, detachment, impersonal society' (p. 13). Money separates the public and domestic life. But at the same time we also try to make the impersonal world meaningful. Again we must emphasize the role of human subjectivity and interpretation in financial-technological practices. Hart's argument relates back to the previous point about the human dimension of finance.

Hart's point about different *forms* of money which each may have different effects when it comes to (de)personalization is also important, since it brings again more variety and plurality in the analysis. Even if the artefact is the same, human interpretation makes a difference. Zelizer (1997) has argued that there is no single money but rather 'multiple monies': people use parallel currencies, and people do not treat cash as entirely impersonal but 'earmark' it, for instance as money for buying a Christmas gift. In this sense, money is not (entirely) impersonal:

> Despite the commonsense idea that 'a dollar is a dollar is a dollar' everywhere we look people are constantly creating different kinds of money. ... [This book] shows how at each step in money's advance, people have reshaped their commercial transactions, introduced new distinctions, invented their own special forms of currency, earmarked money in ways that baffle market theorists, incorporated money into personalized webs of friendship, family relations, interactions with authorities, and forays through shops and businesses. (Zelizer 1997, p. 2)

Money also has a symbolic-psychological aspect: it is related to achievement, status, worry and anxiety, and security (Rose and Orr 2007, p. 743). For instance, people can worry about (a lack of) money and they can save money in order to establish a sense of security (p. 746).

Note, however, that money can only play these many roles and has this multiplicity, *because* it has a universal aspect: its universal exchangeability, already observed by Marx and echoed by many scholars:

> Precisely because it is the universal equivalent, money can serve as the material through which a variety of social relations are expressed. Vespasian was

> essentially right to declare *non olet* (it does not smell) to proud Titus, even when money comes from public urinals. (Fine and Lapavitsas 2000, p. 368 )

Moreover, even if I 'earmark' some money for a specific purpose and thus 'personalize' it, for instance because I received it from a family member as a means to buy a personal present, because of the universal, anonymous *tendency* of this financial medium I might have to make an extra effort to set it apart from 'other' money. Difference is always under pressure when we are dealing with this kind of medium. There are processes of personalization, de-personalization, and maybe re-personalization.

Thus, on the one hand money is impersonal because of its very nature already observed by Marx and Simmel, but on the other hand, if we consider what Zelizer calls money's 'social life' (p. 4), then there is a sense in which it is always again personalized or re-personalized. It has a universal, impersonal side and a particular, personal side. We could say that the social meaning of money is multiple: it is the abstract, universal money economists and philosophers write about, and it therefore *also* has the social and moral meanings Simmel articulated, but it also has and also *can* have many different social meanings. Trade is about detachment from the world, as Marx and Simmel suggested (see Marx on 'the gods of Epicurus in the *intermundia*' [Marx 1867, p. 172]; see also Fine and Lapavitsas 2000, p. 367), but at the same time it is a societal phenomenon through and through. Once again this opens up space for change, at least in principle.

Yet how personal is *global* finance? What meaning and *place* can we give to these specific phenomena, these particular technologies and practices? What if global finance seems to be *all* about impersonal money and impersonal relations? Could we change anything 'there'? And what if we non-experts do not really understand that world? This interpretative exercise is a challenge since it seems that there is a significant gap between on the one hand our daily lives, and on the other hand the impersonal and abstract world of numbers and 'flows': 'We need to understand this virtual world of abstraction in order to make meaningful connection with it from the perspective of our everyday lives' (p. 16). But Hart argues that information can play a role in restoring personal identity since it is now possible to 'attach a lot of information about individuals to transactions at a distance' (Hart 2007, p. 16). Now it is not clear to me why information about individuals would truly *personalize* social relations. My credit card is 'personalized' and my bank account and social media accounts are all 'personal', but this digitalization of identity does not mean the social relations mediated by these technologies are personal. But I think Hart's remark (1) supports my point that if we want to change our lives and society into a more personal direction we need to, among other things, think about changing our (financial) technologies and media and perhaps adapt or change the new electronic information and communication technologies into more 'personal' directions; and (2) reminds us of the anthropological insight that money is about social relations and social meaning, and that it has in this sense an intrinsic *plurality* in it. This means that money itself can play a role in exploring alternative forms of money, that it can be a tool for social-financial change:

> Once we accept that money is a way of keeping track of complex social networks that we each generate, it could take a wide variety of forms compatible with both personal agency and collective forms of association at every level from the local to the global. It is up to us to build them. (Hart 2007, p. 16)

This is all true in principle. Yet it remains a challenge to find and build *global* forms of money, and more generally *global* financial technologies, which at the same time have a strong personal (and perhaps collective or communal) dimension. In the next chapter I will further discuss the issue of social-financial change and explore a number of alternative financial technologies and practices, some of which do seem to have a personal aspect – with or without 'retreat into the local'.

### 7.6. Conclusion: Towards new social-financial practices

It is clear from the preceding pages that the image of global finance created in the previous chapters needs at least some adjustment. First, as Castells already argued, there is the 'global' space of flows but also the 'local' space of places. Financial practices are not located 'nowhere'; as Sassen has shown there are command posts and hubs in international finance and there are also a limited number of global players. Furthermore, people are still located at specific places and communicate with one another. Traders have a global but also a local orientation. Second, finance is a material practice. From anthropology, STS, and social studies of finance we can learn that in global finance there are networks of people and things, of actors and actants (Latour). The world of finance is a world of humans, algorithms, screens, and wires. Trading depends on computers and infrastructures. There is also embodiment, even if traders are oriented towards their screen-world. Global finance has turned out to be less 'virtual' than it seems; it also has a material side to it. It is also not 'virtual' in the sense that global finance has real consequences, for instance for people who depend on food prices. It turns out that the term 'virtual' is rather misleading here. In times when electronic communication has become ubiquitous, we would do better to drop the dualistic virtual/real, digital/non-digital or online/offline categories. Under current conditions created by financial and other ICTs, we and our practices are global and local at the same time, digital and non-digital, virtual and real. Third, there is calculation, but calculation is the result of social and technical arrangements and there is also 'judgement'. Finance – even as mediated by contemporary ICTs – remains a human practice, which means we can re-interpret, re-imagine, and change it. Finally, global finance still leaves room for personal relations and in principle it is possible to re-personalize its practices. This analysis thus shows a more nuanced picture of global finance and opens up space for different social-financial practices.

This means we have to refine the distancing thesis constructed earlier in the book. Contemporary financial technologies in the context of modernity, globalization, and the 'information revolution' contribute to the creation of all kinds of distances, certainly, but this process should not be interpreted only in terms of increasing dematerialization, globalization, and displacement, or in terms of dehumanization and de-personalization. 'Even' electronic financial technologies are linked to material artefacts and infrastructures. This implies that the latter also play an important role in the creation of epistemic, social, and moral distances *and* that they could also help to solve the problem: in principle they could be designed in a different way. Moreover, if, in spite of the term 'global' in 'global finance', the local is still important in financial practices and if 'global' finance is not entirely de-placed (for example, in HFT physical distance matters a lot since the geography of material connections between places matters and firms place their servers at specific places to increase speed), then this means that the geography of finance is far from dead, but is highly relevant to thinking about its ethics.

Indeed, this analysis has implications for thinking about responsibility in global finance and for studying the role of technologies with regard to responsibility. Technological artefacts also shape the conditions under which responsibility is ascribed and exercised and we can now ask more precise questions about how they do this, and what can be done to bring about change in this area. For instance, how do these hybrid 'material'/'immaterial' electronic technologies effectively hide ('screen off') the consequences of financial actions for people at distant locations? Can we re-design them in a way that *shows* these consequences, thus making it easier for traders to act in responsible ways? And how do material artefacts and specific algorithms contribute to a practice that reduces human control and promotes competition? Can we create different things and different codes that help to bring about an alternative financial practice? Moreover, the (re)localization of finance means that centres of financial *power* can be localized, and that people who we think should act responsibly can be *localized*: there are concrete organizations and flesh-and-blood people who can be held accountable and who could help to bring about change.

Change is possible. In line with contemporary philosophy of technology and based on the literature from social studies of technology and finance I have argued that 'technology' is not 'autonomous'. There is no 'iron cage' (Weber), there is no 'system' that is entirely separate from human experience and practice. There is room for interpretation and for taking distance from our practices, including financial practices. Humans are part of these practices. Global financial practices involve networks of human actors and non-human 'actants', both of which are important in bringing financial practices into being, and therefore in changing contemporary finance. Although this does not mean that change is *easy* or that there are no processes of moral, social, and epistemic distancing, it certainly means that there is space for us to do things differently, space to re-think and perhaps re-design financial technologies. This conclusion is also supported by the observation that there is still room for personal relations in contemporary finance, perhaps even in global finance. For instance, we can explore different, less impersonal forms of money. And if money is never something entirely impersonal and abstract but dependent on how people give meaning to it, then perhaps even in the context of contemporary global finance we can try to support processes of re-personalization. To conclude, given the material, localized, human, and personal-cultural aspects of financial technologies, there is certainly room to resist or re-shape the developments and phenomena that I have described and interpreted in terms of distancing.

In the next, last, chapter I further discuss this issue of financial-social change and explore alternative financial technologies and practices. For these purposes I will also need to say more about the issue of power: much of the literature reviewed in this chapter tends to ignore the politics of money and other financial technologies (and theory that could shed light on this issue, e.g., Marx or Foucault).[3] But for thinking about financial-social change this dimension of financial technologies needs to be touched upon.

3 There are exceptions, of course. Barry (2005), for instance, argues that there is a 'politics of calculation'. He argues that measurement has political effects and analyses the – in the UK – famous Hatfield railway accident as a political event, which was about technical things and 'the stubborn fact of metal fatigue' (p. 94) but also about political responses and activities, about inadequacies of a particular form of market organization (the privatization of the railways).

# Chapter 8
# Resistance and alternative financial technologies and practices

## 8.1. Introduction

If we can identify the material/immaterial financial technologies that contribute to processes of distancing, we can try to re-design them in a different way. If we can identify the places and people that are responsible for the development and use of financial technologies, we can hold them accountable. If we can identify the larger social-technological processes that contribute to processes of distancing, we can resist, and/or try to reform and create alternatives. Holding people accountable is necessary in so far as they do not exercise their individual responsibility. *Resistance* is necessary in so far as there are social-economic structures as well as 'meso' and 'micro' forms of power that prevent change. (I already mentioned power and capitalism earlier in this book; in Chapter 2 I used Foucault and in Chapter 3 I briefly commented on Marx). In the 'resistance' discourse the focus is on what we do *not* want, on the indignation we feel in response to current financial systems and practices and their impact on the daily lives of people. Reform of existing systems and the creation of *alternative* social-financial arrangements, by contrast, focuses more on what kind of financial and social future we *do* want and explores how to do things differently. For instance, we may be inspired by what is already going on in so-called 'civil society'. Initiated by non-governmental individuals or organizations, there are already practices that seem to mitigate processes of distancing. And of course we can also try to imagine entirely new practices.

In this last chapter I first review the 'oppositional', 'resistance' discourse, in particular calls for resisting global capitalism. This discourse alerts us to the distance between dominant 'global' players and 'local' people who have far less power, a distance which seems to be increased by means of the new technologies. It also reminds us about the ambiguous role 'machines' play in global capitalism, in particular in their current stage of increased automation: they take over dirty or monotonous jobs, but at the same time they also put people out of work.

Then I turn to the 'reform' and 'alternatives' route and draw attention to what is already going on in terms of financial-social change. I inventory some alternative financial practices people experiment with in civil society and grassroots communities, such as new forms of barter and new forms of currency. I argue that these experiments show that it is possible to do more about the socially and morally problematic aspects of contemporary finance than only blaming particular individuals or 'the system', and that the design of new technologies can play a crucial role in making this change happen. This leads me again to highlighting the close relation between humans and technology. I argue that exploring different ways of doing things does not necessarily mean doing it *without* technology. Rather than a rejection of technology, it seems that what we need is *different* financial technologies which support those new ways of doing things in finance – 'global' or 'local'.

Thus, although in my conclusion I will note that there are limits to these routes of resistance and reform – limits which have to do with the problems and persistence of modern culture – this chapter paints a more optimistic and hopeful picture of the future of finance and technology and encourage us to experiment with concrete alternatives – and indeed pay attention to *what is already happening* in this area. Furthermore, the chapter also pays some attention to the issue of *power* – a problem which is undertheorized in Simmelian discourse about finance and modernity.

### 8.2. Resistance and alternatives

In the previous pages I suggested that there are power structures that create distances and render financial-social change difficult. But which power structures? How exactly do they create distance, and what kind of distance? And if we were to say that these structures need to be resisted, then what or who, exactly, needs to be resisted? There are many ways to answer this question, but an obvious direction to take here is Marx. In Chapter 3 I already visited Marx's analysis of labour under capitalist conditions: this kind of labour creates various kinds of distances. In this section I will further explore views inspired by Marx in order to illustrate the option of an 'oppositional' discourse, a discourse of 'resistance', and to say more about the relation between specific social-economic structures, ICTs, and particular kinds of (social) distancing.

The keyword here is of course 'capital': within Marxian discourse, what needs to be resisted is the power of corporate global capital. Whatever the alternative may be (e.g. communism, economic democracy, etc.), the problem is capital. Now according to one interpretation, 'capital' just *is* 'money machines'. Hart argues that 'the combination of money and machines that we know of as capitalism' (Hart 2001, p. 307) is the source of humanity becoming more unequal and of economic polarization, and that we need to move to a genuinely global and democratic moment in history. He says that we will have to struggle for the value generated by the internet, and need to better understand forms of money and exchange 'with a view to developing financial instruments that serve the interests of each of us and people in general', p. 308). This claim is interesting for the purposes of my book since it puts *financial technologies* at the heart of the problem – and hence the solution. What is the problem according to Hart, and what does it have to do with *distance*? Let me look at his history of what he calls 'the machine revolution' and interpret it in terms of distancing.

Hart thinks a new phase of capitalism – he uses the term 'virtual capitalism' – emerged with the beginning of the futures market. A *new* change is happening and this change has all to do with *distance*:

> Capitalism has become virtual in two main senses: the shift from material production (agriculture and manufacturing) to information services, and the corresponding detachment of the circulation of money from production and trade. This in turn is an aspect of the latest stage of mechanization, the communications revolution of the late 20th century. (Hart 2001, p. 313)

In the previous chapter I criticized the use of the term 'virtual' if it means that contemporary finance is defined by, and reduced to, an immaterial sphere of flows. There is also an important material and indeed 'real' dimension to finance. But here Hart is referring to a distance that I think *is* very relevant and that I also mentioned previously when I discussed for instance the distance between traders and investors and the actual products, people and contexts they trade and invest in: the new technologies make possible distance between, on the one hand, the world of high finance with its 'flows' and its high tech practices, and on the other hand, the world of production, local trade, and flesh-and-blood people who have to live with the consequences of what goes on in the former world. This techno-geographical process has social and political meaning and implications. The new technologies make possible a distance between those who conduct financial transactions 'at the speed of light without regard to borders' (p. 316) and those who do not have these technologies or even live in regions 'stuck' in earlier phases of production. The result is that 'an ever larger proportion of the world economy is controlled by a few firms of western origin but dubious national loyalties' (p. 316). As truly global players (Hart would say 'virtual'), large multinational corporations accumulate power and escape national, territorial control. It seems that individuals, even if they are 'wired', can do very little to resist the power of global corporations or nation states (to the extent that they still have some power and are still relevant). The internet enables new forms of exchange and money, for instance LETS, and in this sense the technology can also be used for empowerment (see also the next section). But there is still inequality, which we can interpret as new social distances created by the new technologies.

Yet if technology is the problem, then it is also part of the solution. Could new technologies alleviate the growing inequalities and reduce social distance? Hart sees 'constructive possibilities' in the new forms of money, including increased personal agency and new ways of more independent, communal living. First, since in the information age exchange is cheap, we can all easily engage in it. It seems that with these kinds of ICTs we can all engage in financial transactions: 'the cheap information contained in bits allows exchange to admit a higher degree of personal agency than before'. (Hart 2001, p. 321). New monies could foster liberty, against the coercion of government and commerce. At least, this is so in principle. In practice I think things are slightly different: we can all engage in exchanges, but the transactions of high finance are still reserved for those who have financial expertise. This means that a gap remains between the high tech world of finance and those of us (the majority) who have to use financial systems *others* have decided upon (e.g., electronic payment systems, banking systems). Usually these kinds of decisions are not taken democratically. Yet Hart is right to point to the *promise* of empowerment linked to contemporary information and communication technologies.

Second, the new technologies promise social change and new forms of 'community'. We can experiment with different ways of living together. Hart admits that current communal systems have problems, but he thinks that this will improve in the future when alternative economic practices will get more sophisticated. At least today 'an arena has been opened up where people can explore different methods of livelihood and co-operation' (Hart 2001, p. 322). The new ICTs can help us to find new forms of community and political association which 'subvert the dominance of a capitalism enjoying the fruits of globalization', thus leading to 'economic democracy':

> The one strategic asset we have is the fast-breaking medium of the internet, and, at a time when society increasingly takes the form of a world market, our efforts at self-emancipation must be focused on the money instruments themselves. For, too often in modern times, the goal of political democracy has been undermined by the absence of any realistic economic counterpart. After the heads and tails of state money and commodity money (Hart, 1986), it is time to make people's money, that is, money in forms reflecting the needs and interests of the people using them. (Hart 2001, p. 322–323)

If we are dominated by global capitalism and if the poor can no longer be protected by nation states; what we need is new forms of money: 'If money is the problem, it is also an indispensable part of the solution'. (p. 323). If we have 'an unfair and unstable system of money-making' (p. 323), this needs to be replaced by 'circuits of exchange based on voluntary association' *and* participation in global markets using electronic systems (Hart 2001, p. 326; see also Hart 2005, p. 13). The new technologies thus create the problem but also give us new possibilities for social change: 'the machine revolution is propelling us towards at least the possibility of forming a democratic human community' and we must use the means of 'improved social connection to combat economic inequality' (Hart 2001, p. 327).

However, Hart's observation that the internet is dominated by the big players (governments and corporations) of this world and that many are excluded from participating in the internet (Hart 2001, p. 328) raises doubts about this vision. Disappointment may invite us to revert to a discourse of 'resistance' against technocapitalism rather than looking for 'alternatives' *within* the financial-technological realm, opposition *against* technology rather than *with* technology. If there is 'domination', then we need opposition: we need to oppose these structures, resist the hegemony of global capitalism. Even new financial initiatives are then to be interpreted as forms of resistance against global capitalism, rather than 'alternatives'.

Let me turn to a more practical issue – food – to further illustrate these two routes: opposition versus creating alternatives, resistance versus reform. On the one hand, new food initiatives can be seen as *oppositional* social movements rather than alternatives, with their focus on present problems rather than on exploring alternatives for the future. On the other hand, they can be interpreted as guides towards a better financial-social future. Let me first say something about the problem these initiatives respond to, before discussing different options to cope with it. I will interpret the problem in terms of *distancing*: I construct the problem in line with Marxian thinking and the *distancing* thesis developed in the previous chapters.

Marx already argued in *Capital* (Vol. 1, 1867) that through capitalist exchange products are abstracted and become impersonal commodities (see also Raynolds 2000). This is also true for food: our 'global' food is impersonal. The persons involved in the production of our food remain invisible, and we 'consumers' are not involved in its production and trade. Through mediation of money and other technologies, products and people become commodified and abstracted. This epistemic distance also creates a 'moral distance' between for instance farmers who produce fruits in 'the South' and consumers who consume fruit in 'the North', and between animals who are bred and slaughtered in distant, invisible places and humans who consume the meat. If these workers, animals, and contexts of living and production are remote from us, then *we do not care*, or at least we care a lot less about them than if they were close to us or if we were involved in their production and trade. As consumers we do not feel responsible because we are ignorant about where the food comes from and because we have no control over what happens to the goods and the animals involved. Consider again the two Aristotelian conditions: in order to ascribe and exercise responsibility, you need to know what you are doing and you need to be in control. Food appears as commodities; in the global supermarket foods do not 'speak' of their context of production and they are alienated from the persons and animals involved. If personal knowledge and personal engagement is lacking, then the animals and their suffering remains invisible. The humans and their exploitation remains invisible. Using Marx and Simmel we can point to the role money and capital play in these processes of food alienation and de-personalization: financial and other technologies make possible and help to maintain these moral and political distances.

Now if the current global financial-social technologies and capitalist social-economic structures help to make possible these forms of distancing, then again an obvious response is to *resist* these technologies and structures. Allen et al. (2003) interpret new food initiatives, especially 'local' production and trade of food, as attempts to *oppose* the global capitalist structures, *resist* the global agrifood system. They borrow the term 'militant particularism' (Harvey) to highlight the oppositional mode of framing problems and solutions. Farmers markets, for instance, are then forms of 'local resistance' (p. 73). Allen et al. also see them as a 'local' movement opposed to 'global' structures. Faced with global 'dominance' the only way forward seems local 'resistance'. They interpret alternative food initiatives in California as movements that

> challenge the time-space distantiation that characterizes the continuing development of the dominant agrifood system. They seek to counter this by building often-local and accountable social relationships – farmer's markets, CSAs, regional foodsheds, short supply chains, fair trade networks – that connect consumers with farmers and that allow consumers to choose in their purchases to support social relations and environmental practices that they value. (Allen et al. 2003, p. 73)

But as the authors recognize, most initiatives actually 'locate themselves carefully within an alternative, rather than oppositional, frame' (p. 74). This shows that a different discourse is possible. Next to the language of 'opposition' and 'resistance' we can also, and need to, discuss *alternatives* to global capitalist networks, alternative forms of financial and economic organization. We still need critique of capitalism and critique of distancing, for sure, but we also need what Leyshon and Lee call 'imagining alternative economic spaces' (2003, p. 7). This approach seems more future-oriented and constructive. Furthermore, as my reading of Hart's vision of social-financial change suggests, we can think about change *against* technology but also *with* technology. If financial technologies are part of the problem, then let us think about how to change them, rather than oppose or 'resist' them. Perhaps new technologies can help us to 'un-distance': to decrease the distancing and disengagement. This is the approach I take in the next section.

## 8.3. Imagining and experimenting with new financial-social practices: How technologies may help us to 'un-distance'

A focus on *problems* of distancing may make us pay less attention to alternative financial-social visions and indeed ongoing attempts to change financial-social practices. There *are* new ideas on how to change finance and there *are* new and emerging financial technologies and new trade and exchange practices, even if these are not always very visible in the media. If we need new forms of money, then let us first look at the new monies that are already emerging. I will inventory a number of new financial technologies and practices, and show that these are never 'mere' technologies and 'mere' financial practices but always have a societal and moral significance and (potential) impact. Can they help to 'un-distance'? What is their social and political 'promise'?

### *8.3.1. Alternative trade and production practices: Fair trade, organic food systems, and farmers markets*

There are alternative trade and production practices that may bridge at least some of the epistemic, social, and moral distance created by modern financial technologies. In trade there have been attempts to set up different systems and networks that reduce the distance between the context of production and the context of consumption. Consider again the domain of food production, consumption and trade. Kirwan has argued that there are spaces of resistance in the agro-food system that allow for strategies to 'produce change in the "modes of connectivity" between the production and consumption of food, generally through reconnecting food to the social, cultural and environmental context of its production'. (Kirwan 2004, p. 395). The Fair Trade network is an example of such a change in the 'mode of connectivity'. As an alternative trade network it does not only insert principles such as fairness into the system, but it also makes consumers more aware of where their food comes from, thus reducing the alienation supported by money and its globalizing, de-territorializing, de-contextualizing, and distancing implications. It thus reduces epistemic distance and thereby moral and political distance, supporting responsible consumption. Although, as Kirwan notes, the Fair Trade network still uses physical structures of the conventional network, it creates and embodies different social and moral relations. It thus responds to the epistemic distance created by financial globalization: through the use of labels and information about where a specific product comes from, the product can be re-territorialized, re-contextualized. Organic food systems which label specific foods as organic and provide information about their origin have this effect.

These alternative trade systems have social but also environmental implications. Raynolds (2000) has argued that 'international organic agriculture and fair trade movements represent important challenges to the ecologically and socially destructive relations that characterize the global agro-food system' and 'seek to create a more sustainable world agro-food system' (Raynolds 2000, p. 297). Thus, the aim of the alternative trade systems is to create new ecological and social relations. But as I interpret Raynolds, there is also an epistemic side to this: these movements also *reveal* something: the ecological and social conditions of food production, the 'relations shrouded by the commodity form' (p. 298). The alternative trade systems do not only *change* the supply chain, for instance, but also reveal it in the first place – to us, the consumers. They also reveal problems in social and ecological problems and needs in 'the South'. This analysis can be used in an oppositional argument (as in Raynolds), but it also suggests a different economic imaginary, which is at the same time a different social imaginary: an alternative economy and an alternative global world in which there are global links between organic systems of production and in which there are fair trade relations.

Yet next to building global epistemic bridges between the context of production and the context of consumption, it is also possible to (re-)localize production and trade itself. In that case the physical distance between producers and consumers is not compensated for or bridged (for example, by providing information about the product) but reduced. Food is produced and sold locally. For instance, there is much less distance between producers and consumers in so-called farmers markets, where there is direct interaction between producers and consumers since the producers sell their own food. This amounts to what Kirwan (following Thorne) calls 'the purposive re-embedding of the exchange process for food in localised social relationships' (Kirwan 2004, p. 407). The farmers market 'facilitates human-level interaction, individual responsibility and mutual endeavour'. It localizes and *re-places* food since consumers can 'directly relate to the place of production' (p. 411). It also localizes and *re-places* trade since consumers directly relate to the place of trade: the food is not traded on invisible 'global' markets and exchanges, but on the local market and in the local village, town, or city. In other words, it is traded and produced in places we may more easily relate to.

Since in these alternative practices consumers, traders, and producers have both more control and more knowledge about what is going on, they can also more easily take on responsibility for what they consume, trade, and produce. It is also easier to hold someone else responsible, to demand a *response* from someone, if that person is not too distant from you. To the extent that these alternative practices manage to bridge or reduce epistemic, social, and moral distance, they replace the vague, invisible 'they' as the subject of finance and trade (the traders, the bankers, the producers 'out there', 'somewhere', 'everywhere') with the 'you' and 'I', perhaps the local or communal 'we'. They bring trade closer to direct social and moral experience and engagement, and therefore give us a real chance to ascribe and exercise responsibility. By changing the conditions under which responsibility is exercised and ascribed, they effectively make possible more responsible production, trade, and consumption practices.

New financial technologies may support such alternative practices, or establish new ones. This is the topic of the next section.

### *8.3.2. Alternative forms of money and finance*

Alternative financial systems and currencies include 'supranational' forms of money, which can either be global and 'stateless' such as Bitcoin or supranational in the sense of a currency used by several states such as the Euro (and to some extent the US dollar), or 'subnational' (Seyfang 2000) forms of money: local currencies and local alternative trade systems. The rationale for these alternatives often rejects the orthodox view of the economy and of money as a neutral medium of exchange separated from the social life and from values. In line with what I have argued in the previous chapters, it replaces the 'dualistic approach' (Seyfang 2000, p. 230) with an alternative view of money, finance, and economy, which connects money firmly with the *social*, with culture, with meaning and values. Money is not neutral, but shapes the social and is shaped by it. This means that alternative monies and currencies, alternative financial systems, and alternative economies are also at the same time alternative ways of living, alternative societies, and embodiments of alternative values. Imagining and experimenting with alternative financial technologies, then, means imagining and experimenting with an alternative society.

Let me inventory and briefly discuss some alternative monies, currencies and financial systems. In each case I will highlight their technological aspects and their relation to social alternatives and social change, with a focus on their potential impact on distancing.

*8.3.2.1. Barter, Global Barter Network, LETS, and time banks*

An alternative to monetary systems which has been around since ancient times is *barter*. In discussions about alternative financial practices, the term may literally mean 'barter', that is, the direct exchange of goods and services without any medium, or it may refer to the use of a medium that is not money in a conventional sense, meaning: not a (supra)national currency such as the US dollar or the euro. The advantage of direct barter is that there is both a personal relationship to the goods and a personal relationship between persons. But media of exchange developed for a reason: they enable much more flexibility. This comes at a price: there is already more distance between persons, and between persons and goods. But there are media and financial 'barter' systems that try to limit the distancing processes by introducing media of exchange that are still different from conventional money. Let us take a closer look at these forms of mediated barter.

Often alternatives forms of exchange are initiated by civil society actors rather than by governments or business. For instance, in the 1990s in Argentina the so-called 'Global Barter Network' or RGT (*Red Global de Trueque*) was developed by people from an environmental NGO in response to inflation and the economic crisis and in order to promote the exchange of goods and services without being restricted by access to money (Pearson 2003). So-called *creditos* were used as currency, a printed currency. The idea was again to create more reciprocal, more 'horizontal' relations and more face-to-face contact (Pearson 2003, p. 219). *Creditos* could therefore be interpreted as a means to mitigate the impersonalizing and distancing effects of conventional money and conventional trade relations. The alternative system promised to render social relations more personal and local.

*Creditos* in printed form are still very much like money, but there are also locally organized and community-based trading systems that do not use money but are still different from direct barter. For example, so-called *LETS* (Local Exchange Trading Systems) do not use notes and coins, but still have a system of credits. The aim is to have a fairer, more social and environmentally friendly economy. This social, community-building aspect is important: the idea is that people enjoy the friendly, community spirit of it. There are many (other) local currencies and community currencies, usually also developed in order to create an alternative economy and better social relations. These are not developed by governments but by local groups. As Seyfang puts it:

> Grassroots activist groups across the world have been working to meet the challenge to create a new kind of money, bringing positive benefits and shifts in our understandings of work, wealth, and sustainability. Their efforts are paying off: currently, alternatives to mainstream money are blossoming the world over. (Seyfang 2001, p. 63)

By itself, it is unsurprising that it is *possible* to have spaces within the economy that are non-monetary; in fact there already *is* such a space. Seyfang (2001) notes that women are often engaged in production and services that are not valued in monetary terms, whereas men tend to have paid employment in the formal economy. But, Seyfang argues, the latter economy depends on the former. She shows that community currencies offer a means to challenge this gender inequality and our ideas of 'work', and help those who work in the non-monetised 'social' economy (mainly women) to gain recognition and recompense (Seyfang 2001, p. 61). Furthermore, LETS and similar alternatives enable local communities to gain some self-reliance (p. 241) and promise to build communities that are more egalitarian and participatory. More generally, barter and alternative currencies such as LETS can bridge what Seyfang calls 'incompatible value regimes' and spaces, for example, commercial and non-commercial spaces.

Ultimately new financial technologies and practices have a political, perhaps even utopian promise: not only a new financial and economic geography, but also a new *society* in which there is far more space for values that get very little space within the conventional market economy. Money, for instance, does not only contain a utopian dimension because it can be transformed into anything (see Chapter 3; see also Dodds 2014, p. 10) and therefore embodies what Simmel called the 'pure tool' – a property which of course captures our imagination; it can also spark off visions of a better society. As Dodds (2014) has argued, money has played a role in many utopian projects. This utopian character of money may have a dangerous side to it (there are of course totalitarian versions of social change which may even be inherent in the very concept of utopia; see also what I said about 'non-place' in Chapter 3: utopian thinking may seek to abolish distance or create infinite distance). But at the same time this connection to utopia also reveals once more the social meaning of money and implies that 'money can be a positive force for change' (Dodds 2014, p. 10). More generally, new financial technologies promise not only new technologies but also new societies, and hence can be a positive force for social change. In the light of the problems of distancing identified earlier in this book, one could rephrase the utopian promise of financial technologies as the promise to bring about *new social and moral geographies*, in which there is less political, social, and economic distance between people, also in terms of gender (see again Seyfang).

Similarly, *time currencies* and *time banks* are meant to have an immediate effect on community and on inclusion/exclusion. Consider, for instance, Time Dollars. People exchange an hour of their time helping someone else on the scheme for an hour's credit which can then be saved, donated, or used to request an hour of someone else's time. A time banker matches requests and keeps a record of the work done (Seyfang 2001, p. 65). Based on her research on time banks in the UK, Seyfang argues that, like LETS, time banks can help to tackle social exclusion. People who join these banks are mainly from socially excluded groups and time banks enable them to practise 'economic citizenship': normally their time is not valued, but now they can exchange time and access services they could normally not afford. At the same time they receive self-esteem and learn new skills. They even feel part of a group, and become involved in wider community activities:

> Time banks provide new channels for mutual support, and give value and recognition to the labour which is normally unpaid and crowded out by the conventional economy, thus challenging the economic imperative of the mainstream market exchange with an alternative based upon a different set of values. (Seyfang 2002)

Thus, these alternative forms of exchange also create different social and moral relations: they reduce the moral and social distance to which mainstream forms of money and exchange contribute and open up new social and moral spaces.

In geography the term 'alternative trading spaces' is used (Hughes 2005) to describe these phenomena and practices. The term can refer to LETS and time banks, but also alternative agro-food networks like farmers markets, fair trade, and (other) labelling schemes. The Slow Food movement should also be mentioned here: whereas originally Slow Food was about regional gastronomic traditions and pleasure, it is now – in the words of Slow Food UK – 'a comprehensive approach to food that recognizes the strong connections between plate, planet, people, politics and culture' including 'where [food] comes from and how our food choices affect the world around us'.[1] What the geographical angle adds to the analysis of these alternative arrangements is that it illuminates the spatial dimension of these alternative forms of exchange and circulation. It shows how alternative forms of trade and exchange intervene in the spatial configurations of 'global' and 'local', and indeed of the social and the political, and thus free up spaces for doing things differently. For instance, geography pays more attention to the scale at which alternative economic practices operate. For example, North has examined the extent to which the ability of alternative spaces to produce economic value might be a problem of scale (North 2005). Do local currencies work best at a local scale? LETS in the UK, for instance, circulate at a very local scale, whereas in Argentina the *creditos* were used at a larger scale. He suggests that as a tool for building alternative economic practices, LETS work best at local scale since otherwise 'the pathologies of the global' are reproduced (North 2005, pp. 230–231).

However, it is questionable if this local-global antinomy is very helpful to understand what goes on here. The spatial interventions of the alternative trading

1 http://www.slowfood.org.uk/about/what-we-do/

and exchange schemes should not be seen only in terms of creating something 'local' as opposed to something 'global' which remains unchanged. The 'global' is always also re-configured. For instance, in contrast to farmers markets, fair trade is not only 'global' or 'local', but both at the same time. The supply chain is not physically shortened, as in the face-to-face transactions between producers and consumers on the farmers markets, but the consumers' knowledge moves down to where the food originates. The antinomy global-local is not very useful to describe this new mode of connectivity. Moreover, even the farmers markets have a global aspect: they are not isolated phenomena but can be interpreted as belonging to a global alternative culture.

Furthermore, it is important to emphasize, given my focus on financial technologies, that many of these alternative trade and exchange systems are made possible by *technologies* and are mediated by these technologies in specific ways. For instance, with contemporary information and communication technologies, it becomes much easier to do things such as labelling and tracking of food, or time banking. Computers and networks enable us to (re-)connect production and consumption, 'North' and 'South'. Sometimes old technologies are used, such as the spatial-material arrangements of a market. But all the current alternatives are made possible by means of technology (new or old), that is, by means of specific mediations that bring about a new mode of connectivity between producers and consumers.

One approach, therefore, is to interpret these alternatives as *alternative socio-technical systems*, a term we may borrow from STS. For instance, Leyshon and Lee have argued that actor-network theory (ANT) can be used to re-imagine, rather than only resist, the current system: the network metaphor can help to point out 'where potential exists for new forms of alliances, social formations and economic geographies first to take root, then to become established, and finally to flower and bloom' (Leyshon and Lee 2003, p. 12; also quoted by Hughes 2005, p. 500). The idea is that by revealing new networks and assemblages of humans and artefacts, we can try to show new, hopeful possibilities and develop an alternative vision, a vision of a different financial-economic world.[2] But whether or not ANT in particular is the best approach here, studying the interconnections between technologies and the social can help us to understand the role of *technology* in how current economic spaces work and how alternative trading spaces and financial spaces can come into being. We should not look at technology and the social in isolation, but analyse them together in order to study its social and ethical implications. Anthropological methods such as ethnography (as also used by STS) can help here.

---

2 Of course, there is always a danger that the alternative vision becomes itself commodified and drawn into consumerism. For example, some argue that Fair Trade has been commodified; others argue that the mainstreaming of Fair Trade into commercial distribution channels has not eroded its counter-hegemonic character (Low and Davenport 2006).

How does this work? Perhaps technology's role is most clear in the case of less material (after the previous chapter we should avoid 'non-material'), electronic forms of money and currency. Here we meet *Bitcoin* again. Let me draw on Fletcher (2013) and other recent research in order to show how an anthropological, social-technological approach to new financial practices can shed light on the role of technology: not only its role as a tool, but also its role as an agent of social-financial change.

*8.3.2.2. Bitcoin*

What is the social and political 'promise' of Bitcoin technology? How could it change society? First, Bitcoin is not neutral when it comes to power relations; it is 'political' in this sense. As noted before, Bitcoin is decentralized, and its cryptographic and peer-to-peer design allows Bitcoin users to operate outside of central issuing authorities and therefore beyond the scope of conventional regulation (Fletcher 2013). Instead, Bitcoin relies on what we may call a 'horizontal' form of trust and social relations. In Fletcher's words:

> Trust between economic actors sustains market systems reliant on bitcoin. Having neither an intrinsic value nor any representational aspect, its value is subject to market forces and user confidence. ... In effect, bitcoin's value is largely a representation of the market's faith in it. (Fletcher 2013, p. 9)

Because of its electronic, 'immaterial' dimension, which is embedded in the design of the technology, Bitcoin challenges the nation state's monopoly over money and the more 'vertical' form of trust embedded in national currencies. Governments seek to retain control but lawmakers have problems defining its properties: what kind of money or currency is it? (Fletcher 2013, p. 10). The result is that to some extent Bitcoin escapes control by the nation state and projects a model of societal relations that is not so much based on 'vertical' trust of citizens in the government, but rather peer-to-peer 'horizontal' trust and 'horizontal' social relations. In that sense, it promises to reduce distance: it promises to replace the 'vertical' distance between individuals and states with an electronically mediated form of proximity and direct exchange between individuals. Let me emphasize again that it is Bitcoin's *technological* features that create this social and political significance. Its peer-to-peer design is both 'technological' and 'social'. It creates a new system or network of relations between people and things and thus reconfigures social and political space.

Second, this also means that Bitcoin – again because of its technological characteristics – is 'political' in the sense that it propels a societal vision, a social and political ideology, perhaps a utopia. The more precise nature of this vision and ideology can be described 'from the outside', as I have done so far in this book, but we can also use ethnographical research to achieve a view 'from the inside': the 'Bitcoin community', the 'Bitcoin culture'. This is what Fletcher does in his ethnographical research: he interviews people from the Bitcoin community in the

United States (in particular the Central Florida area) and asks them about their political and social views, in particular about Bitcoin and social change. Many of them support some version of social change, even if they are not all 'anarchists and liberals' but also and mainly 'technologically savvy private business owners and tech industry entrepreneurs' (Fletcher 2013, p. 55). In the community Fletcher found, the predominant ideological direction seems to be anti-state and libertarian. One Bitcoin user ('Jack') sees Bitcoin as a way to achieve denationalization and freedom:

> With bitcoin Jack saw something that could serve as an alternative to fiat currency, something he believed to be truly transformative. … 'It's freedom', Jack exclaims, 'No one can tell you what to do with or who to send it to'. … 'It is an "economic anarchist's" wet dream!' (Fletcher 2013, pp. 36–37)

Given the context of Fletcher's Bitcoin community, the kind of freedom meant here is probably not that of nineteenth century European anarchism but rather twentieth century 'free market' thinking, for instance the views of Hayek, who advocated private monies (Hayek 1974):

> The dream of a viable decentralized currency, once little more than the hypothetical discourse of F.A. Hayek (1974) has been realized in the early 21st century. Bitcoin has become the first sustainable digital, decentralized, and denationalized currency to resonate among consumers. The past three years have seen bitcoin rise from obscurity to become a highly sought after commodity. (Fletcher 2013, p. 58)

I suggest that Fletcher would have found very different ideologies in different Bitcoin communities, in different parts of the world and in different cultures. There is no such thing as 'the' Bitcoin community. All the same, Fletcher shows how people in the Bitcoin community he studied have a 'desire to see a change in today's social and economic organization' and use the technology to interact with the world around them (pp. 58–59) and to contribute to that change. These Bitcoin users really believe in the transformative power of the currency and the technology. For them, the technology is obviously not a 'mere' technology; it is something that has transformative social power and that may realize a political dream (whatever that dream is).

In order to further reveal the promise and power of Bitcoin as an agent of social change, it is also worth mentioning the analysis by Maurer et al. of the aspirations of Bitcoin adherents in terms of what the authors call the 'semiotic' value of Bitcoin:

> The point is not whether Bitcoin 'works' as a currency, but what it promises: solidity, materiality, stability, anonymity, and, strangely, community. Indeed, …

> Bitcoin combines a practical materialism with a politics of community and trust that puts the code front and center. (Maurer et al. 2013, p. 263)

Maurer et al. argue that the novelty of Bitcoin lies in the reflection it provides on the sociality of money. Bitcoin was meant to 'free people from the tyranny of middlemen' such as banks and money shippers (Lyons in Maurer et al. 2013, p. 265) and to keep transactions private, without surveillance or censorship. It replaces both the personal relations of face-to-face trade (personal trust) and the impersonal relations of trade via intermediaries (trust in intermediaries) and impersonal trust in state authorities with – in the inventor's own words – 'a system for electronic transactions without relying on trust' (Nakamoto 2008, p. 8). Or rather, as Maurer et al. argue, there is trust in the algorithm.

I doubt if Nakamoto and Maurer et al. are right about what they say about trust. First, if there is trust in the algorithm then this means that, in contrast to what Nakamoto wants, the system still relies on trust. Second, even if there is trust in the algorithm, that does not mean that the system no longer requires trust between people. As suggested before (see also again Fletcher), the currency still depends on trust people have in others using the currency. 'Having faith in' the currency really depends on the social – in this case on impersonal but 'horizontal' relations. Peer-to-peer networks and practices depend on peer-to-peer trust. The nature of money is always both social and technological, it cannot be *merely* technological, as Nakamoto's utopian vision suggests. Yet it is interesting to see how a financial technology is firmly connected with views about social and political relations. Financial technologies should not be seen as belonging to a separate 'financial' or 'technological' world. They intervene in the social, economic, and political or, better, they are an intrinsic part of it.

This is also true for currencies in *virtual worlds* and in *online games*, to which I now turn. At first sight, they seem to belong to a separate, 'virtual' realm. But such a view is misleading since it blinds us to the real nature and consequences of these electronically mediated phenomena. Let me explain this.

*8.3.2.3. Money in games and virtual worlds*

There are monies and currencies that could be seen as even more 'virtual' than electronic currencies such as Bitcoin: monies and currencies used in computer-mediated 'virtual' worlds and 'online' multiplayer games. Some may even say they are 'fictional', and that they therefore do not deserve any treatment in a book on financial technologies. And if they are 'virtual' and 'fictional', then it may seem that they cannot raise any ethical or social issues. Then they seem irrelevant with regard to our 'real' concerns and problems.

However, this view is misguided. As I already argued previously, any so-called 'virtual' practice and technology is very real indeed. It is real since it is a human and truly social practice which takes place in the online/offline world in which we live, and it is real in its epistemic, social, and moral significance. We should reject any suggestion that there is a clear separation between, on the one hand, a 'real', 'physical' world and, on the other hand, a 'virtual', 'un-real' fictional world. In the new world that has come into being today, both spheres have merged – if they were ever separate at all. (And even if one took the view that previously they were separate, one could say that there is a sense in which today 'virtual' currencies become more real and 'real' currencies become more virtual: both kinds of transactions now mainly take place by electronic means and hence their ontological status is at least similar, if not identical.) If this is true, then it means that we should treat currencies 'in' 'virtual' worlds and games in the same way as we treat currencies and monies in the so-called 'real' world, and study their social and political significance. Let me further discuss their nature (the question concerning their 'virtual' nature) and their potential as 'alternative' social-technological arrangements.

Online games, in particular massively multiplayer online role playing games (MMORPGs) such as World of Warcraft, and virtual worlds such as Second Life all have currencies and in-game markets. For example, Second Life has 'Linden dollars' and World of Warcraft has 'gold'. This means not only that there are in-game economies and exchanges, but also that these in-game economies interface with world economies. For instance, Linden dollars are exchanged with currencies such as the US dollar. Game items are exchanged for 'real' money. People are willing to pay 'real' money to buy virtual money and virtual goods. Virtual assets thus get real-world value. Some people actually earn a wage by accumulating and selling virtual goods or game currency (see, for instance, Heeks 2009). It turns out then that the 'virtual' currencies are very much like any other currencies. People use them to purchase goods, people buy and sell them, thus exchanging them for other currencies. They do not only resemble money, they just *are* money. This is also the view of Castronova, an economist who studied the business and culture of online games and virtual worlds: 'It is frankly impossible to deny that the gold pieces of fantasy worlds are money, just like the money in your pocket. They are sustained by exactly the same social mechanisms and perform exactly the same functions' (Castronova 2005, p. 151).

Thus, these monies are not only real because they can be exchanged for 'real' money; they are also real *money* because they have the same function and the same social nature. Both 'internal' and 'external' monies are real. What matters is their function and social acceptance (see also my discussion about money and social ontology), which seems independent of support by national governments. Castronova explains:

> If everyone thinks a certain piece of money has value, they will treat it as a valuable thing, and therefore it will have value. When I hand someone a worthless old scrap of paper that says '$1', she will give me something valuable – a Coke – in return, because of the institution of money. And it is the institution, the patterns of behaviour, that actually gives the dollar bill its value; the government has little to do with it. (Castronova 2005, p. 102)

Like Simmel, Castronova suggests that the function and nature of money is not dependent on its specific material form; what matters is its function as a medium and its *social* nature. The fact that these virtual currencies appear within a computer-mediated world does not change their status as money. What makes or breaks their status is their functioning as a means of exchange and their ability to gain social acceptance and trust. Castronova argues that for instance a 'silver piece' in the world of his character 'Sabert' 'is not merely like money, it is money. It is genuine, actual money, defined as "any commodity or token that is generally accepted as means of payment"' Castronova 2005, p. 47).[3] Once it is money, no one can turn it into 'un-money just by imagining it to be so' (p. 48). More generally, the worlds of games and virtual worlds are not to be situated outside but fully inside the sphere of 'ordinary human affairs' (p. 2), and this is also true for its economic and financial phenomena and practices. Again money in its different forms is shown to have a significant social and human nature; like other (financial) technologies it is not a mere instrument.

Moreover, the emergence of these new financial economies has effects on global finance. This includes problematic effects. For instance, virtual currencies can be used for criminal activities. Bronk et al. argue that today the landscape of financial transactions includes 'systems explicitly designed for financial opacity', allowing transactions that are meant to be hidden (Bronk et al. 2012, p. 129). Thus, the new technologies enable new and alternative monies and exchanges, but also create epistemic distance. This opacity and complexity is used by criminals, and if we ignore what goes on in so-called 'virtual' worlds because we think it is 'un-real', we do so at our peril. In addition, there are also ethical and political issues with the digital labour related to these games and virtual worlds, for instance political economy issues. It seems that in these real economies there is an invisible labour force at work: players-workers who do not appear in the 'official' economy, but who nevertheless do work for others and get paid for it. But do they get paid enough? What are the circumstances in which they have to work? Who defends their rights and interests? Nakamura, inspired by Miller, suggests we must follow the money to its less glamorous and virtual place such as the 'gold farmer sweatshops', for instance in China (Nakamura 2009, p. 131; see also Zhang & Fung 2013). Here we meet again the alienation and distancing Marx identified: the work may be less dirty and dusty than in the 'real' factories, but it is boring and workers are indifferent towards their products. The digital sweatshop worker 'does not develop freely his physical and mental energy but mortifies his body and ruins his mind' (Marx 1844a, p. 74; see also Chapter 3). And with the end of

3 In his recent book on virtual money, Castronova provides a more elaborate analysis of what money is, and surprisingly adds the 'joy' function of money. I doubt if this is a *necessary* part of the definition of money, but all the same it is interesting again to see that also in this respect he makes no difference between 'real' and 'virtual' and argues that virtual money seems to fulfil this function: although 'you can't run your hands through piles of virtual gold', 'owning virtual money makes people happy'. (Castronova 2014, p. 121)

the separation of work and play it seems that capitalism has fully conquered the lifeworld: it might appear that there is only one sphere left, one in which there is only labour and capital, and in which alienation and exploitation can flourish.

But there are also people who see opportunities for positive social transformation. The new games, worlds, financial technologies, and economies also express political hopes and utopian desires. They can be used for exploring alternative economic and financial arrangements. Some may desire what Dibbell calls 'ludocapitalism' (2006, p. 299): why not celebrate the end of the separation of work and play, rather than regret it? The new games and worlds then promise to realize the 'ludologist' dream of play which invades the whole of life (p. 295).[4] Others may hope for other social and political outcomes, for instance community, collective ownership and socialism. In games there are all kinds of ways of dealing with property and there are democratic and authoritarian versions of virtual worlds (see, for instance, Lastowka & Hunter 2004). These social and political possibilities are not limited to contemporary arrangements; many games and worlds refer to historical forms. For instance, observing players of MMORPPGs, Brignall has argued that there is a 'tribalism' in these worlds: 'Online communities offer individuals the ability to find and interact with people who share a common identity, often amplifying tribalistic behaviour' (Brignall 2008, p. 111) – with all its negative and positive aspects. With regard to the latter, he observes that players of World of Warcraft (WoW) especially enjoy the socializing, even *prefer* the socializing in WoW to 'offline' socializing:

> Reported reasons included a feeling of strong friendship, group unity, the ability to role-play an alternate identity, hanging out with people who had similar likes, social anonymity, and the ability to ignore disliked people. A majority of hard-core players reported living in isolated areas with a limited potential for finding friends. A common response was it was easier to meet people in WOW. (Brignall 2008, p. 114–115)

Whether or not this picture of players of WoW (and players of MMORPPGs) is accurate, it is clear that these games offer alternative social environments, which may well have negative aspects to them but also, among other things, create opportunities to reduce social distance and explore alternative societal arrangements. In that sense, the electronic technologies – including the financial technologies – are, or at least can be, agents of social change. They may help us to question, if not to change, our financial and social institutions.

4 'Ludology' can refer to a particular approach within game studies, but here it appears to refer to a normative position which, inspired by the work of Huizinga and others, emphasizes how important play is in human culture (and thus also in computer games) and celebrates play and the *homo ludens* (the 'playing human').

*8.3.2.4. Microcredits and web-based peer-to-peer lending*

Another financial innovation is the institution of microcredit and, more generally, microfinance. The idea of microcredits is to give small loans to individuals and communities, usually in developing countries: the aim is to alleviate poverty and enhance the capabilities of poor people in these countries. For example, a small loan can give someone without access to regular credit channels a chance to start up a small business, to buy essential household equipment, or to study. In this way, microcredits can empower individuals (especially women) and communities. In its modern form the institution has existed since the 1970s, but recently the information revolution is shaping new possibilities. Today electronic technologies support and transform the institution of microcredit by enabling more 'horizontal' forms of lending. Today the internet can be used for so-called 'peer-to-peer lending': websites and platforms such as Kiva and Zidisha enable lenders to give direct loans to people in developing countries with only a small or no interest rate. This practice can be seen as an alternative to more 'vertical' types of social-financial relations between on the one hand large, rich, and powerful financial institutions (e.g,. banks) and governments and on the other hand individuals without much power and money. It also gives individual lenders the freedom to use their own judgement. Today this is combined with a 'social media' aspect: for instance on Kiva I can decide to whom I want to lend by looking at pictures and information of several candidates. This 'social media' aspect may have problematic sides, for example: do most lenders choose to lend to the person with the greatest need, or do they choose the person who is culturally close to them or who can present her case in an attractive way? Moreover, success in these 'social media' type of peer-to-peer lending networks – microcredit or not – may very much depend on what Lin et al. call the borrower's 'social capital' (Lin et al. 2009). Consider also the practice of crowdfunding: on the internet it is easier to raise monetary contributions from many people if you already have enough 'social capital'.

In any case, it is clear that electronic technologies have an impact on the practice of lending in a way that seems to bridge social and economic distance. Both the Web 2.0 technologies and peer-to-peer lending are 'based on mutual and social exchanges between people instead of centrally controlled communications and relationships' (Ashta & Assadi 2009). Yet Ashta & Assadi argue that in practice the relations are not very direct: the role of the intermediary remains important and, for instance, in the case of Kiva, parties do not need to trust each other but rather trust the intermediary (Ashta & Assadi 2009), in this case Kiva.org and indirectly the 'Field Partner', the local microfinance institutions Kiva works with. The poor are not directly involved in the communication. But in principle the Web 2.0 technologies 'promise' and make possible such directness and hence promise to redress the distancing problems: they *can* personalize and bridge the epistemic and social distance maintained by mainstream lending institutions of global finance, and already do so to some extent.

If it works, microcredit has implications for responsibility. Individuals (for instance in 'the North') have a closer and more personal relation to the people they lend money to (in 'the South), know more about what they are doing (epistemic condition: they know what happens to their money and to whom they give it) and have more control over what they are doing (control condition: they can judge and decide how much to give to whom). This means that via the new technologies, individual citizens such as you and I can actually exercise a much more direct form of responsible action with regard to alleviating poverty, rather than delegating that responsibility to large and distant organizations and/or governments. And recipients of the micro loans can, in principle, also know much more about where the money comes from and what the deal is, which may help *them* to act responsibly (e.g. pay back the loan, but also do something meaningful with the money, something that really empowers them and enhances their capabilities). The new technologies can, again *in principle*, decrease the distance between lenders and borrowers and *establish* a relationship between them which supports responsible and meaningful financial action.

Note that this argument for a more personal, 'closer' form of credit practice does not imply that therefore banks and governments are *not* responsible. Governments and commercial banks also have a role to play since (1) not all social problems can be solved by means of microloans, (2) banks can also set up microcredit systems, and (3) the existence of microcredit institutions and technologies should not be used as an excuse for not changing 'macro' financial practices and institutions – and financial, economic, and social policy. Prior and Argandoña (2009) argue that financial institutions have social responsibility to help create a financial system that makes credit instruments available to the greatest number of citizens, that such institutions have a *social* function, and that if they operate in a developing country they have special responsibilities that arise from that environment. Providing microfinance services to more people, expanding their outreach, is a way to exercise that responsibility.

#### *8.3.2.5. The euro*

Money can also bridge political, economic, and social distance between nation states. Let me have a brief look at a supra-national currency which is usually not categorized as 'alternative' but which is relatively new and which is clearly linked to visions of, and controversies about, spatial, social, and political change: the euro.

Historically, the very idea of a single European currency already goes together with a normative, political vision. For instance, Jacques Delors, president of the European Commission from 1985 to 1995, wanted a monetary union but also common economic policies and co-operation on social issues. One of the founding fathers of the European Union, Jean Monnet, wanted a federation of European states in order to bring and maintain peace, prosperity, and social development in post-war Europe. Already during the war, on 5 August 1943, Jean Monnet gave his vision on the future of Europe, which contained an idea about the relation between space and political organization and which also mentioned social development:

> The countries of Europe are too small to guarantee their peoples the prosperity that modern conditions make possible and consequently necessary. They need larger markets. ... Prosperity for the States of Europe and the social developments that must go with it will only be possible if they form a federation or a 'European entity' that makes them into a common economic unit. (Monnet 1943)

When the euro was launched in 1999, the single currency was seen as a means for creating a single market, but also for closer cooperation between member states. To put it in terms of distance, the euro was meant to contribute to decreasing the distance between member states and to the formation, development, and consolidation of a larger, supranational political-spatial entity.

In 2002 euro cash was introduced: bank coins and cash. Interestingly, on its website the European Central Bank emphasizes the importance of cash and its 'unique features', which include the following: it is a fast and cheap instrument for retail transactions, it is 'inclusive' since people with no access to bank accounts or electronic forms of payments can still make payments, and it enables people to keep a close check on their spending.[5] Thus, even in the electronic age, a non-electronic form of money was chosen for its 'unique features', and one of these features directly refers to a social policy aim. Discussing the euro, then, means discussing politics and society, not 'just' finance.

This becomes especially clear if we consider the discourse about the so-called 'crisis' of the euro (partly also in response to the economic crisis of the past years). According to critics, the euro should be a means of exchange that helps to establish a single European trade zone, but should only do that and not support more European integration. Some even want to abolish the euro; they want each country to follow their own monetary course. For others, the euro is one of the roads towards a more integrated Europe: financial and economic, but also political and social. The former are accused of egoism and lack of solidarity; the latter are seen as advocating a new bureaucratic monster. Thus, the 'monetary' issue is in fact entangled with a political issue concerning the desirability of further European integration (or not) and this issue is often framed in moral and political terms: it is presented as a choice for or against solidarity, for or against cooperation, for or against including particular countries, and so on. At stake is not only the future of the euro, but also – in this case – the future of Europe. José Manuel Barroso, President of the European Commission from 2004 to 2014, said in a speech at the University of Düsseldorf: 'Do we Europeans stick together in dealing with these internal and external challenges or do we try to resist separately? Are we stronger in numbers by pooling our resources or should we be doing things by ourselves?'.[6]

Again we observe that money and currency are never neutral – neither morally nor politically. The euro may be seen as more orthodox and perhaps far less spectacular than some of the alternatives discussed previously in this chapter, but

5 https://www.ecb.europa.eu/euro/intro/html/index.en.html

6 http://europa.eu/rapid/press-release_SPEECH-14-289_en.htm

it becomes clear that discussions about 'financial' change are also and perhaps mainly about societal change. Here the key issue is how much distance in terms of economic and social policies there should be between European nation states, indeed if the European political space should become more or less united and integrated. More generally, the discussions about the euro raise the question of how today, in the context of globalization, finance and politics should be organized. The future of money is also the future of our societies and our political institutions. Money can create distance and it can bring people together. Answering the question 'Which money do we need?' is answering the question of how we want to live together with whom in what kind of political-social spaces.

*8.3.2.6. Slow Money*

A movement worth looking at in the context of financial change and that is highly relevant to the distancing issues discussed in the previous chapters is Slow Money. Taking its name from the Slow Food movement (see Section 8.3.1.), it opposes 'fast money'. One of the Slow Money Principles reads: 'There is such a thing as money that is too fast, companies that are too big, finance that is too complex. Therefore, we must slow our money down – not all of it, of course, but enough to matter'.[7] This is a problem of 'speed' but also a problem of 'place', or more specifically: of de-placing and of disconnection or *distancing*. Tasch, initiator of the movement and former venture capitalist, sees 'economics disconnected from bioregions and communities, markets disconnected from places, wealth disconnected from health'. (Tasch 2008, p. 5) This also reminds us of the problem of the distance between investors and investment, which I mentioned previously. In Tasch's words: 'Investors don't know anymore where their money goes and more and more, they want to see an impact for what they give in their own lives and own communities'.[8] This suggests that the epistemic distancing I identified is not only a problem noted by people outside the world of business and finance; apparently some investors *also* want to shorten the distance between investors and their investments. They feel a responsibility towards their community and perhaps towards the environment. Instead of 'fast money', therefore, the movement promotes steering capital to small food enterprises, organic farms and local food systems linked with integral and organic funds, municipal bonds, and local stock exchanges.[9] These are what Tasch calls new forms of 'financial intermediation':

> What is needed is a new form of financial intermediation. Intermediation that favors a large number of small, independent, local-first food enterprises over a small number of large, consolidated, global-first food enterprises. Intermediation

7 http://slowmoney.org/principles

8 http://www.justmeans.com/blogs/slow-money-a-new-movement-for-social-enterprise

9 http://en.wikipedia.org/wiki/Slow_Money

> that is oriented around nurture, rather than acceleration, extraction, and export. (Tasch 2008, p. 80)

'Slow Money' then means bringing the rhythm of the economy in line with the slow metabolism of nature and humans (p. 76) and going 'back down to earth' in the following sense:

> We need to steer money, it its primary applications, that is, in those functions toward which it is deployed in the making of money, toward life, toward enterprises that enhance the quality of life, that preserve and restore fertility, biodiversity, and the health of bioregions and communities and the households that live in them, and away from enterprises that degrade quality in the name of quantity. [This is the] use of money, of investment capital as an antidote to the disease of excessive quantification. (p. 93)

Consider again Simmel's view about the quantification of life and society: the Slow Money movement explicitly opposes quantification here. It also seeks to (re-) connect capital markets to *place*. What this means becomes clear if we turn to the Slow Money Principles again, which claim that we must 'connect investors to the places where they live, creating vital relationships and new sources of capital for small food enterprises'. It asks: 'What would the world be like if we invested 50% of our assets within 50 miles of where we live?' Thus, the Slow Money movement combines the ideology of the Slow Food movement and organic farming with a programme that explicitly looks for different financial intermediations and tries to re-place, re-localize and de-quantify finance, in particular by decreasing the distances between investment, food production, and places where people live. In this vision the fast, big, and complex world of global finance must be replaced with slow, small, and transparent capital and investment structures in order to arrive at an economy and a society that shortens the distance between investors and their investments, and that is better in terms of food, health, ecology, and preservation and restoration of the environment. The responsibility promoted by this view goes far beyond so-called 'corporate social responsibility', which Tasch rejects as 'a dangerous distortion of business principles' (p. 51), and instead proposes 'socially responsible and sustainable investing directly in individual small food enterprises near where we live'.[10]

In terms of technology, Slow Money seeks to replace the speed and the gigabytes (p. 4), the 'gigantic technology' (p. 102) and 'gigantic machines that destroy nature' (p. 252), and the large industrial organizations where 'robots run the show' (p. 73) with 'millions of small acts and restraints, conditioned by small fidelities, skills, and desires' (p. 102). Although this shows that Tasch recognizes the important role of technology in the economy, these remarks seem to suggest a discourse of resistance against 'technology' rather than thinking about *alternative*

10 https://slowmoney.org/other-resources

financial technologies. But Tasch's view can easily be revised and re-interpreted in this direction. His thinking is orientated towards exploring alternatives, and the 'small' and 'slow' economy he proposes does not need a rejection of 'technology'; instead, it needs thinking about *which* alternative technologies need to be combined with *which* kinds of 'fidelities, skills, and desires' and *how* to reduce the various kinds of distancing observed. Since they are so much bound up with the human and with society, thinking about 'slow' finance and 'slow' living must include thinking about 'slow' technologies.[11] Again the main idea is to focus on alternatives: If we do not want the current money machines, let us create new ones.

*8.3.2.7. The gift?*

Finally, given the distancing problems discussed in the previous chapters, is it better to use no money at all? Or more radical: is it even better to reject barter, and instead resort to the *gift*, which at first sight seems to involve *no* intermediation? This depends on what 'gift' means. There are two possibilities. Either it is an anonymous, 'pure' gift with no strings attached which, as Laidlaw, Mauss, and other anthropologists have argued creates no connections between persons (Laidlaw 2000). In that case, the gift does not seem to solve the distancing problems identified, if it exists at all.[12] Or it is a gift that is *not* pure and thus involves some kind of exchange and interests, and *does* establish relations. In other words, it is a gift in the common meaning of the term, which is always to some extent a medium of exchange. In the latter case, the gift can play a role, like money, as a mediator in social relations. In fact, it already does so: we use the (not entirely pure) gift in contexts of personal relations, for instance when we give something to a friend or a partner.

However, could the gift be used outside these personal contexts? Could it help to re-personalize what is now regarded as 'other domains'? It is imaginable, for instance, that it could play a role in employment relations. To the extent that they are modern, today such relations are largely shaped by money and numbers. Perhaps work could sometimes be 'given' and 'received'. Then it would no longer be 'labour' that is exchanged for a wage, and this might then mitigate Marxian alienation effects. However, it is not clear to me that the gift could ever *replace* money. I speculate that even if a wider application of the gift is capable of making possible more personal and better social and moral relations (thereby contributing to an alternative modernity or perhaps to less modernity), it would be difficult not

11 In fact, there are already 'Slow Tech' and 'Slow Technology' movements and manifestos, but these are not necessarily in line with the view I develop here. I hope to discuss them elsewhere.

12 Derrida has argued that the gift is an impossibility since the absence of reciprocity would mean that both recipient and giver should not recognize the gift as a gift (to avoid feelings of obligation or interests such as satisfaction). Therefore, as soon as there is a gift it becomes part of a cycle of exchange and thereby ceases to be a gift (Derrida 1994; see also a summary of his view in Venkatesan 2011).

to use money at all. The reasons for this have to do with tradition and habit, but also with how well money performs its function as medium. As Simmel shows, money has a long history in our civilization and is very much bound up with its media and its institutions. There are already so many monetary technologies and institutions that provide 'bridge' functions and we – as societies and cultures – became habituated to them. Habits are difficult to change. Like other technologies, money has become part of 'us', that is, part of what it is to be human and to live in societies. In the course of civilization we became so addicted to money's flexibility and its role as a nearly perfect means (see again Simmel's pure tool; money is the perfect technology), and indeed to the kind of 'bridging' relations and social-spatial forms money makes possible, that it is very likely that we cannot do without a similar kind of medium of exchange. That being said, perhaps it is recommendable to keep open some space for the (non-pure) gift in human relations and society, as *one* of the means to cope with distancing and disengagement problems.

This topic needs more discussion, but let me end this section by observing that the non-pure gift not only already has a place in our practices, but that these practices are also changing as a result of the new ICTs. Today electronic technologies are creating new monetary practices *and* new gift practices. For example, I can use social media to send someone a 'gift card' or (in the context of social media) I can give someone 'likes'. We then can ask: does this electronic item and this medium really 'bridge' the distance or maintain it (or even create more)? What does the medium do to the relationship? The point is, again, not if this kind of gift is 'virtual' or not. It is real, and it has social and moral consequences. But which consequences? The key question with regard to distancing, both in the case of money and in the case of artefacts used for a non-pure gift ('virtual' or not), is *what kind of* media create or bridge what kinds of distances in which kind of practices, situations and contexts. What kind of technologies and media can support less distant relations and more responsible action in these practices, situations and contexts? And what is the role *financial* technologies and media can play in this respect?

## 8.4. Conclusion: The future of financial technologies

In the previous sections I have given an overview of candidates for 'alternative financial technologies and practices'. There are of course more examples, for instance recent movements for financial reform (e.g. Positive Money in the UK, which aims to 'democratise money and banking so that it works for society and not against it'[13]) and historical alternatives that played and play a significant role in a European context such as credit unions and social banking (Benedikter 2011); this is only a selection. The point was to illustrate *that* there are alternatives and to explore how some of them may intervene in the social-spatial and moral-spatial

13 http://www.positivemoney.org/about/

structures of global and local finance and how they promise to re-shape not only finance but also society and politics. Admittedly, it was not always clear how 'alternative' these financial technologies and institutions were. Some of them may well support a Simmelian evolution of money and media in the direction of 'ever more insubstantial versions', re-forming and revealing money 'as pure information' (Hart 2005, pp. 13–14) – thereby perhaps also re-shaping *humans* and human relations in a more global, insubstantial, and informational direction. Maybe this is the course electronic currencies such as Bitcoin take. Others – in fact, most of the alternatives discussed – attempt to re-connect, re-localize, re-materialize, and re-personalize global and local finance. Note, however, that the latter option is not necessarily 'anti-technological' or anti-ICT at all. On the contrary, a practice such as online peer-to-peer lending shows that ICTs can also be used to try to bridge social and moral distances. Yet it remains difficult to assess to what extent such attempts contribute to social and moral change: at first sight, the technologies seem successful in decreasing distance, but on closer examination it is not clear if they keep their 'promise'. This may happen because the technologies and media are not 'purely' applied and therefore they cannot function as for instance a pure electronic medium which creates direct and personal relations (for example, there are still too many intermediaries in the case of peer-to-peer lending) or because it seems to lie in the nature of electronic ICTs that at the same time they also *create* or maintain distance (e.g. Bitcoin, electronic lending platforms). What also happens is that a technology that seems at first sight a distancing technology, for example Bitcoin, is *also* used and revealed as an instrument for the building of social relations and community.

It turns out that financial technologies have many faces. Ultimately, this is due to the intrinsic human and social character all technologies and media have. The faces of financial technologies and their ambiguities are *our* multiple human faces and *our* human ambiguities. They are not something external to the human or external to the social. Money is about us, about social relations, and about society. As I have stressed again and again in this chapter, exploring alternative monies and new financial technologies means exploring alternative social and human possibilities and shaping new social-spatial configurations. The future of finance, as entangled with the future of electronic technologies and the social and moral relations they help to shape, is also the future of humanity and the future of our societies, cultures, and civilizations.

This is why we should not leave decisions about money and financial technologies, and the design of these instruments, to financial experts alone. In my conclusion I will not only summarize the discussion about financial technologies and distance and further reflect on social-financial change, but also recommend a democratization of financial-technological-social innovation.

# Chapter 9
# Conclusion

## 9.1. Financial technologies and distancing: Conclusions for the moral geography of global finance and for social change

In this book I have identified and discussed various forms of distancing: the epistemic, social, and moral distancing Simmel, Marx, and other critics of modernity observed, which seem to worsen due to new, electronic financial technologies and under conditions of globalization. Phenomena such as high-frequency trading and Bitcoin show how current financial technologies, in particular electronic ICTs used in contemporary finance, make possible the growth of a global sphere of 'flows' which is increasingly run by computers, a world of numbers and other abstract entities which give us only abstract information about what happens to others at distant locations and which seem to jeopardize the flourishing of personal social relations and the exercise of moral responsibility. It turns out that we live in a 'global city' rather than a 'global village', with all the advantages and disadvantages of a large modern city such as intensive trade, substantial and fast flows of money, good transportation and communication connections, technological and cultural innovation, frequent impersonal interaction, anonymity, and lack of care for the common good. And even within this 'global city', the 'city of finance' appears to alienate itself from the daily lives and reality of those who cannot 'make money with money'. There is a social and political distance between, on the one hand, the world of high finance and high tech and, on the other hand, the world of those who have little idea about where their money comes from or goes to, but who have to 'go with the flow'. Thus, it appears that financial technologies have contributed to, and are continuously contributing to, processes that illustrate the best and the worst of modernity.

Today it seems that we find ourselves approaching the peak of these developments. The new ICTs enable us to bridge distances faster than ever before. Unless regulation slows our communications down, the speed of light will be the only remaining speed limit. This is true for finance, but also for other areas where electronic ICTs and infrastructures are used to intensify information and communication. Yet I have also argued that there is a cost to this gain in information and communication, a cost which I suggest is created by all kinds of ICTs and that *all* of us bear in one way or another: we successfully bridge or shrink physical distances, but at the same time we create epistemic, social, and moral distances. Our screens give us all the information, all the numbers we want and through electronic platforms and 'social' media we can communicate with people at the other end of the world. But are we really in touch with the reality 'at the other end of the screen'? Do we really *know* what we are doing when we act remotely? How engaged are we when we use these electronic technologies? How 'personal' and 'social' are our contacts? And can we assume full moral responsibility for what we do if we do not directly experience the remote consequences of our deeds and if we delegate so many tasks to machines? All this seems especially problematic in the case of global finance, given that new forms of money make possible a form of trade which is incredibly fast and remote, anonymous, abstract and quantified. Where are the humans of flesh and blood who might gain or suffer from these transactions? Where are the humans responsible for these transactions and their implications for the economy and for society? Are we drowning in these currents of numbers and oceans of data? Are we cogs in a gigantic 'money machine'? Who owns the money machines, the automata of global finance? Does *anyone* control them? Are the only survivors the 'machines' and maybe those who own them and feed off them as parasites? Is this the new moral and political geography of global finance? Put in these terms, our predicament is bleak and hopeless.

However, in the course of this book I have not only constructed but also *nuanced* this picture of contemporary global finance and its relation to financial technologies in several ways.

First, modernity is not all to blame for distancing and disengagement effects. In Chapters 2 and 3 I have argued that while these processes of distancing might well be peaking now or will do so in the near future, their origin and development must be situated much earlier in the history of our civilization as they accompany the history and development of ICTs and society. Modernity may have intensified the distancing effects, but in ancient times financial and other technologies already made possible more epistemic, social, and moral distance. By using Simmel I have offered a broader perspective on these issues which goes beyond criticism of modernity. To fully understand the epistemic, social, moral, and political implications of financial technologies we need to widen our scope to the histories of cultures, technologies, and civilizations.

Second, as I have argued in Chapters 3 and 7, the 'distancing thesis' should not be read as assuming or promoting a deterministic story about the unavoidable development of technology, finance, and society in the direction of total alienation and dehumanization. The 'rollercoaster' image that is suggested by the Simmelian analysis of money and financial technologies has its limitations, although it does a good job at communicating the *speed* of the changes that are taking place and the *urgency* and societal relevance of the issues discussed in this book. Similarly, the 'Singularity' concept is misleading in this respect since it is far too deterministic and one-dimensional, even if it attends us to the issue of lack of understanding: for most of us, and indeed even for many experts, the point where we no longer comprehend what exactly goes on in global finance and what its technologies mean for our society has already been passed. Equally, the 'machine' metaphor has its limitations if it is taken to refer to 'mechanistic' processes. There is still room for human freedom. We live in a world of algorithms and computers, but we are not simply or certainly not *only* cogs in the wheel or bits in the streams of information. While the new technologies may now appear to us as being 'external' to us, as being 'alien' entities which 'take over', this is only part of the picture. 'Machines' are always deeply connected to humanity; they are as much 'internal' as they are 'external' to the human. While 'money machines' play a central role in contemporary global finance, its practices are and remain 'human' in several senses: as I have argued in Chapter 7 there is still room for human judgement, and technologies and technological practices are always fundamentally human since humans design, create, and use them. This means that humans can change things; the 'machines' are not all-powerful since they are not totally external 'things' in the first place. The metaphor of the industrial machine (think of Chaplin's film *Modern Times*, for instance) and its wheels and cogs is not adequate here; the metaphor of the cyborg or McLuhan's 'extension' imagery, for instance, might work better since they are not only based on a better and deeper understanding of contemporary ICTs and media, but also on the premise that humans and technologies/media are and always have been entangled.

Third, the image of 'the global space of flows' describes only part of what is happening. There is certainly an important immaterial and de-territorialized and de-placed aspect to contemporary global finance. I also think Simmel was right when he pointed out that in the course of the history of money (and I extended that to the history of *financial technologies*), we can observe a process of dematerialization. Yet in this book I have also paid attention to more concrete spatial, geographical implications of the new financial technologies. Drawing on more empirically oriented work in STS, social studies of finance, geography, and other social sciences I have argued in Chapter 7 that place and materiality still play a role in contemporary global finance. For instance, there are financial centres and financial 'hubs', and trading practices turn out to involve material tools, specific places, and infrastructure. This means that we *can* give answers to questions about 'where' 'those people' are as far as traders and financial experts are concerned. Much more difficult, perhaps, is to say where the people are who invest (capital owners) and where the people are who are affected by contemporary global finance (their interests, their social and economic situation, etc.). Either way, it is significant that *work* has to be done to reveal these places and these material realities. The immaterial, placeless and perhaps 'spiritual' face of global technofinance (and global ICT) seems to be much *more* visible. We can see the numbers and the statistics. What remains hidden, or at least is much less visible, includes the remote workers and their labour which global finance needs in order to function, the geographically situated social and economic inequalities, incapabilities, disempowerment, and injustices it creates, and the place-bound material artefacts and infrastructures used to sustain its clean upfront 'sphere of flows'. In addition, if the material side of global finance and its technologies is more or less hidden, this means that as a society we do not usually realize quite how much global finance – and indeed all of us when it comes to electronic ICTs – depends on this material basis. It means, for instance, that we are generally not aware of vulnerabilities related to the material infrastructure of financial ICTs.

Traders, experts, the media, and the general public focus on what 'the market' does and what 'the numbers' tell us. In the meantime, our economy is dependent on a number of servers, wires, and other equipment that are located at very specific places and have a much less ghostly, much more concrete, material, and earthly form than suggested by the 'digital', 'flow' faces of global finance that reach our screens. When something happens to this technological infrastructure (serious malfunctioning, crash, or attack), we may well pay a high price for our epistemic alienation from these artefacts, a form of distancing which so far might have appeared as a mere academic issue.

Finally, it turns out that the 'distancing thesis' I initially constructed is too one-sided and misleading in so far as it suggests that current financial ICTs always and necessarily have these distancing effects, that there is no room for responsibility and for other relations, and that we cannot change their epistemic, social, and moral influence. This (interpretation of) the thesis is problematic in at least the following three senses, which suggest different implications for social change and different potential responses to the problem of 'distancing'.

First, as I suggested in the previous pages there is room for responsibility and different social and epistemic relations *in spite of* what current financial technologies do to our knowledge, social relations, and responsibility. If the thesis is interpreted in this way, then we need to look for change outside the area of electronic ICTs, perhaps even outside the area of money, finance, and the economy. Then we need to make sure that the current technologies do not dominate us and that they no longer or do not further 'colonize' our lifeworld (to use a Habermasian phrase). Then 'alternative financial practices' probably boils down to either reinforcing a strict modern separation between spheres (the financial and technological spheres versus the sphere of the lifeworld) or going back to earlier financial technologies and practices, perhaps even to barter and gifts. But this interpretation takes 'technology', here 'electronic financial technologies', as given and understands them as having only one possible kind of influence.

Second, however, it is also possible to see the new technologies as epistemically, socially, and morally *ambiguous*. There may be all the effects and problems mentioned, but at the same time the very technologies under consideration may *themselves* 'promise' change. Maybe, to paraphrase Heidegger again, the 'danger' might also 'save' us (see my remark in Chapter 4). I think the basis for this hope lies again in the recognition that technologies are not something entirely external to the human (as Heidegger's own use of the term 'technology' might suggest) but are rooted in, and entangled with, human cognition, human activity, and human values. The epistemic and normative possibilities that lie in contemporary ICTs, including financial ICTs, mirror, express, or maybe even realize possibilities that lie in humanity and in human culture. More social interconnectivity and more global moral, social, political, and spiritual awareness is one such possibility; more social and moral distance is another one. As McLuhan suggested perhaps the new technologies also open up possibilities for more 'integral' ways of living, for *more* engagement rather than less, perhaps for new forms of proximity. Why not bringing about change by enhancing and strengthening *that* side of the technologies and the financial and other practices which they shape? For instance, could we re-shape and re-orient current financial technologies in such a way that they support, rather than threaten, global and local social and moral communities? As Simmel's analysis suggests, money was always already socially and morally ambiguous: it has enabled people to connect and bridge, but at the same time it also made relations less personal. Similarly, current ICTs such as electronic currencies may well create all kinds of distances, but we might want to think about how we can strengthen their 'bridging' function. For instance, we could try to strengthen and use their peer-to-peer aspect, which has at least the *potential* to de-couple financial relations from nation state governments, bureaucracies and large corporations, and experiment with alternative global or local financial-social communities. We could also use current ICTs to create more epistemic transparency, in global finance and elsewhere. But if technologies have always inherent social and moral effects, 'strengthening' some aspects rather than others cannot only be done by *using* technologies differently; rather, we also have to *change* them. We can try to re-shape and re-orient technologies in the normative direction we desire and agree on. This idea of 're-shaping' and 're-orienting' technology brings me to the third and related option.

Third, if we do not take 'technology' as given, then this opens up the possibility of changing the technologies. If it is true that, as I suggested in this book, financial technologies play such an important role in society, then changing these technologies may contribute to positive social and moral change. As I have proposed and shown in Chapter 8, we can conceive of, explore, and experiment with, new financial and social practices. I have shown how each of the proposed alternative forms of trade, money, and finance contain a social and political 'promise' and might help us to mitigate the distancing made possible by mainstream financial technologies and institutions.

Combining the second and third option, then, we can call for change – financial, technological, social, and moral change – and think about what is required for this change to happen. In geographical and communication-technological terms, we need places of innovation and we need more connections. First, there is no doubt that this is possible. If we live in a 'global city', then that also means there is room for hotbeds of technological, financial, social, cultural, and artistic innovation (and cross-fertilization between these). We can think about how existing and future initiatives on alternative finance and alternative financial technologies can be supported, at a global level and at other levels. We can support connections between these levels. We can create and search for 'alternative' places and practices that take place 'outside' the financial centres, but also explore how existing places of financial-technological innovation 'within' the financial world can be re-built into places where technological and financial innovation goes hand in hand with social and moral innovation. We can also explore ways to connect the world of finance to the world of citizens. Second, therefore, this program requires, among other things, that the philosophy and ethics of financial technologies be connected to practices of research and innovation in the field of ICT. If the analysis presented in this book is adequate, then it is not enough that 'business' and 'finance' change into more responsible directions; it is also both necessary and recommendable that designers and developers of financial technologies take up more moral and social responsibility for their practice. Moreover, if financial technologies are really as important as I suggested, then we should not leave decisions about them to a small number of people; instead we must *democratize* the use, design, and regulation of financial technologies. Let me end this book by reflecting on the implications of my discussion for (1) academic work on financial technologies, in particular in philosophy and ethics of financial technologies, (2) responsible and democratic research and innovation in the development of financial technologies, in particular financial ICTs, and (3) our lives.

## 9.2. Conclusions for philosophy of technology, the need for responsible and democratic research and innovation in the development of financial ICTs, and the challenge to change our lives

One of the aims of this book was to address a lacuna within philosophy of technology. As I observed in my introduction, so far philosophy of technology has not paid much attention to financial technologies. For example, as I indicated in chapter 6 there are only a few papers on the ethics of automation in finance. A systematic framework for understanding and evaluating financial technologies is still lacking; there is still plenty of room for philosophers of technology and ethicists to work on this issue. The same is true for work on financial technologies carried out under the banner of responsible innovation. As Muniesa and Lenglet argue, in contrast to other disciplines such as biotechnology, there is no ethical framework that is to guide innovation in the sector. Indeed, the very notion of responsible innovation is rare in finance (Muniesa and Lenglet 2013).

With this book I have started to fill these gaps. While it does not (and was never meant to) deliver a straightforward list of guidelines, criteria or principles for ethics of financial technologies or responsible innovation in finance, it offers philosophical and sociological reflections on financial technologies – in particular ICTs – that highlight potentially problematic aspects of these technologies and reflects on possible alternatives. By showing how financial ICTs make possible various modes of 'distancing' and by discussing their implications for responsibility, it has made a contribution to thinking about the ethical and social implications of global finance and financial technologies. Rather than immediately focusing on ethical principles (which is the dominant approach in ethics of finance) or rather than going straight into discussions of specific contemporary financial products or new forms of currency, the book focused on *understanding* financial technologies. It started with the history of financial technologies, located its analysis within thinking about technology, and has connected its discussions to reflections on modern culture and society. This approach was inspired by an important insight from social studies of science and technology and more 'socially' and 'culturally' oriented philosophy of technology: technologies do not stand alone (as means to ends, as instruments) but are entangled with society and with human values and culture. Technologies are not only tools people use to do things, but they also shape *how* people do things, shape the practices and ultimately the society in which they have been developed and in which they are used. This is also true for contemporary financial ICTs and the financial innovations they made possible.

Therefore, if we are to think of how to guide and regulate innovation in finance in a more ethical and responsible way, it does not suffice to offer ethical principles or methods for moral reasoning. While we must applaud *any* attempt to render global finance more ethical, this book suggests that if we really want more responsible financial practices, we need also need to attend to (1) the use and development of technologies and (2) to the social and cultural context in which these technologies are used and developed. What does this imply for further research?

First, in order to further understand and evaluate the use and development of financial technologies, we can use the approach and methodological tools offered by empirically oriented *philosophy of technology* and by thinking and practice in the field of *research and responsible innovation.* It is important to reflect on financial technologies pro-actively, before and during their development, rather than waiting until they are used, and to open up processes of technological research, design, and development in a way that democratizes them. Today there is a gap between on the one hand the context in which financial technologies are developed and on the other hand democratic processes and spaces for ethical reflection and public engagement. As Von Schomberg remarks in a paper on responsible innovation, modern technology has been 'democratized in its use and privatized in its production' (Von Schomberg 2013, p. 53). This is also true for financial technologies (think about the monies we use), and some are not even democratized in their use (for example, the algorithms of high-frequency trading). This economic, social, moral, and political fragmentation needs to be studied and changed. Unless we transform the moral and political geography of research and innovation, ethical evaluation, socially responsible action, and adequate regulation and policy will always come too late: after the technology is already in use and after it already has changed our world and has had its epistemic, social, and moral impact. Therefore, we need to think about how to make research and innovation in the development of financial ICTs more responsible and more democratic. For this purpose, we can learn from existing work on responsible innovation, ethics, and democracy,[1] even if it is not directly concerned with *financial* technologies. How exactly can we create an innovation environment that promotes the Aristotelian conditions of responsibility I discussed in this book and that enable financial experts to act responsibly? Which concrete institutional and technological tools can we use to facilitate the development of technologies that create less rather than more epistemic, social, and moral distance? How can we involve stakeholders in

1 See for instance work by Richard Owen or Bernd Stahl.

processes of financial innovation and make it more democratic? How can we make citizens aware of the societal importance of financial-technological changes?

Second, this 'societal' aspect is crucial. Thinking about financial technologies also means thinking about the relation between financial technologies and society. This was an integral part of the approach taken in this book. It became clear that changing financial technologies requires not only changing technologies and innovation, but also changing our societies and culture. It does not suffice to demand from people working in the finance sector to behave in a more ethical way, to analyse the financial risks of specific financial products or financial and economic models, or to try to render research and innovation in finance more responsible by focusing on technological design and development processes only. If financial technologies are a crucial part of society, then that means that their development and use is deeply connected to the very structures and infrastructures of our globalized and modernized world, to its histories, meanings, and practices, and to the problems and new opportunities that arise within that world. Any work on ethics and on responsible innovation in finance therefore also needs to make connections to this wider societal and cultural context. Any attempt to understand financial technologies that does not take into account broader historical, social, and cultural contexts will not only be incomplete; it will also offer too little insight into the problems and therefore turn out to be incapable of finding better and more integral solutions. For example, this book suggests that understanding *modernity* and globalization is crucial for interpreting the nature and impact of money and other financial technologies. Further research on financial technologies could benefit from making more connections with research on modernity, globalization, and related concepts (consider, for example, well-known work by Giddens, Bauman and Beck in sociology).

This also means that the connection between financial technologies and wider social-cultural processes and contexts is a double-edged sword. On the one hand, as I emphasized here and in the previous chapter it opens up possibilities for social change. There is no need for fatalism. On the other hand, with Marx, Weber, Simmel, and Heidegger we should keep in mind that there are limits to social and cultural change and therefor to change in financial practices. While technologies do neither determine us nor preclude *any* change, their entanglement with social and cultural developments and conditions such as modernity and indeed developments that span the history of civilization (see my interpretation of Simmel's views about money in terms of a history of distancing) means that financial-social change is never a matter of will, design, and action alone. Resistance and reform in finance are possible, but necessarily take place within a material, social, and cultural horizon. Change can only happen within a space that emerged in the course of history, that changes also without our explicit efforts at change, and that is never entirely within our control. For example, we can try to change processes of distancing by means of new technologies and alternative financial practices, but to a significant extent we do not control the results of our (technological) actions. Even our thinking about technology and finance is dependent on this moving ground, on this dynamic organic structure, which is partly shaped by humans but at the same time also conditions us. For instance, it is difficult for us to think about a world without money or with entirely different media, or to create a discourse about finance that weaves together finance, technology, and the social in totally new ways. We might be able to stretch and move the boundaries, but we cannot remove them. There are already technologies, institutions, languages, and places; change will always necessarily involve ex-change with what is already there and what shapes what we do, think, and desire.

Finally therefore, given these limits to change and control, the ubiquity of financial technologies and the many and complex ways in which they shape our world, it would be naive to think that changes in the financial sector, academic research and reflection, and policy-driven research and responsible innovation are sufficient when it comes to bringing about significant change in finance and developing attractive, real, and successful alternatives. First, even if we were to achieve more responsible innovation and design, we never have full control over social-technological change and innovations. Even if we *want* to change things and create a new technology or initiate a better process of innovation, there is always going to be a distance between our ideal and what actually happens, and often unexpected and unintended things happen that are difficult to predict. This is certainly the case with technologies and social relations in a global and complex world. We cannot simply 'design' and bring about a new world, and attempts to do so are usually rather dangerous and violent. We may rather want to influence the *growth* of new financial-social practices. Second, we also need to involve other modes of thinking and doing than those predominantly practiced in finance, academia, and policy. We need to pool more kinds of creativity and know-how to explore different financial languages (and therefore different ways to think about finance) and to find more responsible and truly innovative ways of doing finance. For example, *artistic work* relevant to financial technologies and society can help us to reflect on how we think about money and other financial technologies and explore alternative ways of talking about finance, of doing finance and of living together. We need to think about how to connect artistic practice to (other) financial, technological, and social innovation processes. We should not only turn to the *politics* of finance and economics but also experiment with its *poetics*. Third, change should not only come from 'above' or from 'somewhere else'. We can also conclude from this book that fundamental change in finance is only possible if it happens in conjunction with social innovations and *changes in the way we live*, and this has consequences for our responsibility for change at individual and communal level.

Let me explain the latter claim. Financial technologies and our lives are interdependent. On the one hand, our lives are shaped by the tools and media we use. There is a sense in which we *are* our tools, and this is also true for money and other financial technologies. The money we use, for instance, shapes how we (literally) deal with others. I suggested in this book that money and the *form* of money matters to how we think and how we relate to others. Experimenting with new forms of money, then, is experimenting with new ways of life and new ways of living together. This is why it is important to think not only about use but also about innovation. Thus, starting with changing the technologies and their use in the places where they are usually designed, developed, used, and regulated is one option. On the other hand, I stressed that our technologies do not exist in isolation from the social and from our lives, but are part of it. I have argued that we are not determined by our technologies; we can – within limits – change things, improve things. Now this 'we' includes individuals and communities. If we are *not* just cogs in the money machine, then we cannot just delegate responsibility for changing finance and financial technologies to 'them': to others, far removed from us and our daily lives. If we want to explore alternative financial technologies, institutions, and practices, then this means that our professional and private lives should also be examined and changed – even if also in this area radical change may be difficult or impossible. Changing financial technologies is changing ourselves, our relations to others, and our world. If we want to 'un-distance' or 'slow down', then this is not only a task for politicians, experts in finance, or people who develop financial and other technologies; it is also our personal challenge and a challenge in the context of personal relations and local communities. Technologies and institutions are not alien things and structures, even if under modern conditions they often appear like that; they depend on humans and human lives as much as we depend on them. This limits us but also gives us space for human freedom and responsibility, at all levels. Many of the alternatives discussed in the previous chapter were not initiated and developed 'top-down', by governments or by large corporations; they were 'bottom-up' innovations at grassroots level. Change can also start with personal reflection and initiative and with communal and collective attempts to do things differently. Alternative financial technologies and practices involve new habits, new skills, new kinds of relations, and new rituals and institutions. These habits, skills, relations, rituals, and institutions are all 'near' to us. If change is going to happen, we better make sure that we, as citizens and communities, have a say in it and have some influence on it. It is at least partly up to us – persons, groups, and communities – to take up responsibility and participate in creating and experimenting with new technological-social forms. By trying out alternatives, we may experience what works to reduce the distance and find new, better forms of mediation and engagement – in global finance and in our lives.

# References


Allen, John & Pryke, Michael. 1999. Money Cultures after Georg Simmel: Mobility, Movement, and Identity. *Environment and Planning D: Society and Space* 17(1): 51–68.

Allen, Patricia, FitzSimmons, Margaret, Goodman, Michael, & Warner, Keith. 2003. Shifting Plates in the Agrifood Landscape: The Tectonics of Alternative Agrifood Initiatives in California. *Journal of Rural Studies* 19: 61–75.

Andreau, Jean. 1999. *Banking and business in the Roman world.* Cambridge: Cambridge University Press.

Arendt, Hannah. 1958. *The Human Condition.* Chicago: University of Chicago Press, 1998.

Aristotle. 2001. *Nicomachean Ethics* (trans. W.D. Ross). In *The Basic Works of Aristotle* (ed. R. McKeon). New York: The Modern Library.

Ashta, Arvind & Assadi, Djamchid. 2009. Do Social Cause and Social Technology Meet? Impact of Web 2.0 Technologies on Peer-to-Peer Lending Transactions. *Cahiers du CEREN* 29: 177–192. Retrieved from: http://www.veecus.com/photo_presse/press_social%20cause%20and%20technology.pdf

Augustine. *Confessions* (trans. R.S. Pine-Coffin). London: Penguin Books, 1961.

Barry, Andrew and Slater, Don (eds). 2005. *The Technological Economy.* London: Routledge.

Barry, Andrew. 2005. 'The Anti-Political Economy'. In Barry, Andrew and Slater, Don (eds). 2005. *The Technological Economy.* London: Routledge.

Bartelson, Jens. 2000. Three Concepts of Globalization, *International Sociology* 15: 180–196.

Benedikter, Roland. 2011. European Answers to the Financial Crisis: Social Banking and Social Finance. *Spice Digest*, Spring 2011. Retrieved from http://fsi.stanford.edu/sites/default/files/social_banking.pdf

Beunza, Daniel & Stark, David. 2004. Tools of the Trade: The Socio-Technology of Arbitrage in a Wall Street Trading Room. *Industrial and Corporate Change* 13(2): 369–400.

Beunza, Daniel, Hardie, Iain, & MacKenzie, Donald. 2006. A Price is a Social Thing: Towards a Material Sociology of Arbitrage. *Organization Studies* 27(5): 721–745.

Bijker, Wiebe, Hughes, Thomas & Pinch, Trevor (eds). 1987. *The Social Construction of Technological Systems: New Directions in the Sociology and History of Technology*. Cambridge MA/London: MIT Press.

Boatright, John R. 1999. *Ethics in Finance*. Oxford: Blackwell.

Boatright, John R. (ed.) 2010. *Finance Ethics: Critical Issues in Financial Theory and Practice.* Hoboken, New Jersey: John Wiley & Sons.

Bolter, Jay David & Grusin, Richard. 1999. *Remediation: Understanding New Media*. Cambridge, MA: MIT Press.
Borgmann, Albert. 1984. *Technology and the Character of Contemporary Life: A Philosophical Inquiry*. Chicago: University of Chicago Press.
Brey, Philip. 1998. 'Space-Shaping Technologies and the Geographical Disembedding of Place'. In Light, A. & Smith, B. (eds). *Philosophy and Geography Vol. III: Philosophies of Place*. Rowman & Littlefield.
Brignall III, Thomas. 2008. 'Guild Life in the World of Warcraft: Online Gaming Tribalism'. In Adams, Tyrone L. and Smith, Stephen A. (eds). *Electronic Tribes: The Virtual Worlds of Geeks, Gamers, Shamans, and Scammers* (pp. 110–123). Austin: University of Texas Press.
Bronk, Christopher, Monk, Cody, and Villasenor, John. 2012. *Survival: Global Politics and Strategy* 54(2): 129–142.
Callon, Michel (ed.). 1998. *The Laws of the Markets*. Oxford: Blackwell.
Callon, Michel & Muniesa, Fabian. 2005. Peripheral Vision: Economic Markets as Calculative Collective Devices. *Organization Studies* 26(8): 1229–1250.
Callon, Michel, Millo, Yuval, & Muniesa, Fabian (eds). 2007. *Market Devices*. Oxford: Blackwell.
Castells, Manuel. 1996. *The Information Age – Economy, Society and Culture, Volume 1: The Rise of the Network Society*. Oxford: Blackwell.
Castronova, Edward. 2005. *Synthetic Worlds: The Business and Culture of Online Games*. Chicago/London: Chicago University Press.
Castronova, Edward. 2014. *Wildcat Currency: How the Virtual Money Revolution is Transforming the Economy*. New Haven, CT: Yale University Press.
Clark, Gordon, Thrift, Nigel & Tickell, Adam. 2004. Performing Finance: The Industry, the Media, and its Image. *Review of International Political Economy* 11(2): 289–310.
Clark, Gordon L. 2005. Money Flows Like Mercury: The Geography of Global Finance. *Geografiska Annaler: Series B, Human Geography* 87(2): 99–112.
Coeckelbergh, Mark. 2012a. Technology as Skill and Activity: Revisiting the Problem of Alienation. *Techne* 16(3): 208–230.
Coeckelbergh, Mark. 2012b. *Growing Moral Relations: Critique of Moral Status Ascription*. Basingstoke/New York: Palgrave Macmillan.
Coeckelbergh, Mark. 2013a. 'Information Technology, Moral Anxiety, and the Implosion of the Public Sphere: A Preliminary Discussion of the McLuhanian Problem of Responsibility'. In Van Den Eede, Y., Bauwens, J., Beyl, J., Van den Bossche, M., and K. Verstrynge (eds). Proceedings of 'McLuhan's Philosophy of Media – Centennial Conference'. Contact Forum.
Coeckelbergh, Mark. 2013b. *Human Being @ Risk: Enhancement, Technology, and the Evaluation of Vulnerability Transformations*. Dordrecht/New York: Springer.
Crotty, James. 2009. Structural Causes of the Global Financial Crisis: A Critical Assessment of the 'New Financial Architecture'. *Cambridge Journal of Economics* 33: 563–580.

Davies, William & McGoey, Linsey. 2012. Rationalities of Ignorance: On Financial Crisis and the Ambivalence of Neo-Liberal Epistemology. *Economy and Society* 41(1): 64–83.

Davis, Michael, Kumiega, Andrew, & Van Vliet, Ben. 2013. Ethics, Finance, and Automation: A Preliminary Survey of Problems in High Frequency Trading. *Science and Engineering Ethics* 19(3): 851–874.

Derrida, Jacques. 1994. *Given Time, vol. 1: Counterfeit Money*. (trans. Peggy Kamuf). Chicago: University of Chicago Press.

Dibbell, Julian. 2006. *Play Money: Or, How I Quit my Day Job and Made Millions Trading Virtual Loot*. New York: Basic Books.

Dobson, John. 1997. *Finance Ethics: The Rationality of Virtue*. Lanham, Maryland: Rowman & Littlefield.

Dodd, N. 2012. Simmel's Perfect Money: Fiction, Socialism and Utopia in *The Philosophy of Money*. *Theory, Culture & Society* 29(7–8): 146–176.

Dood, Nigel. 2014. *The Social Life of Money*. Princeton, NJ: Princeton University Press.

Dreyfus, Hubert L. 2001. *On the Internet*. London: Routledge.

Dreyfus, Hubert L. & Kelly, Sean Dorrance. 2011. *All Things Shining: Reading the Western Classics to Find Meaning in a Secular Age*. New York: Free Press.

Ess, Charles. 2009. Floridi's Philosophy of Information and Information Ethics: Current Perspectives, Future Directions. *The Information Society* 25: 159–168.

Farag, Shawki M. 2009. The Accounting Profession in Egypt: Its Origin and Development. *The International Journal of Accounting* 44(4): 403–414.

Ferguson, Niall. 2008. *The Ascent of Money: A Financial History of the World*. Allen Lane/Penguin Books.

Fine, Ben & Lapavitsas, Costas. 2000. Markets and Money in Social Theory: What Role for Economics? *Economy and Society* 29(3): 357–382.

Fletcher, Justin Harrison. 2013. *Currency in Transition: An Ethnographic Inquiry of Bitcoin Adherents*. (MA thesis, University of Central Florida, Orlando, Florida).

Floridi, Luciano & Sanders, J.W. 2004. On the Morality of Artificial Agents. *Minds and Machine* 14, 349–379.

Floridi, Luciano. 2007. A Look into the Future Impact of ICT on Our Lives. *The Information Society: An International Journal* 23(1): 59–64.

Floridi, Luciano. 2008. Information Ethics: A Reappraisal. *Ethics and Information Technology* 10(2–3): 189–204.

Floridi, Luciano. 2009. Against Digital Ontology. *Synthese* 168(1): 151–178.

Floridi, Luciano. 2013. *The Ethics of Information*. Oxford: Oxford University Press.

Foodwatch. 2011. *The Hunger-Makers: How Deutsche Bank, Goldman Sachs and Other Financial Institutions Are Speculating With Food at the Expense of the Poorest*. Berlin: Thilo Bode/Foodwatch.

Foucault, Michel. 1975. *Discipline and Punish: The Birth of the Prison* (trans. Alan Sheridan). New York: Vintage Books, 1995.

Foucault, Michel. 1982. 'Technologies of the Self'. In Martin, Luther H., Gutman, Huck, & Hutton, Patrick H. (eds). *Technologies of the Self: A Seminar with Michel Foucault*, (pp. 16–49). Amherst: University of Massachusetts Press, 1988.

Franssen, Maarten, Lokhorst, Gert-Jan, & Van de Poel, Ibo. 2009. 'Philosophy of Technology'. In *Stanford Encyclopedia of Philosophy*. Retrieved from http://plato.stanford.edu/entries/technology/#DevEthTec

Fuchs, Christian. 2011. 'The Contemporary World Wide Web: Social Medium or New Space of Accumulation?' In Winseck, Dwayne & and Jin, Dal Yong (eds). *The Political Economies of Media: The Transformation of the Global Media Industries,* (pp. 201–220). London: Bloomsbury.

Galbraith, John Kenneth. 1975. *Money: Whence It Came, Where It Went*. Boston: Houghton Mifflin.

Giddens, Anthony. 1990. *The Consequences of Modernity*. Stanford, CA: Stanford University Press.

Gilbert, Emily. 2005. Common Cents: Situating Money in Time and Place. *Economy & Society* 34(3): 357–388.

Graeber, David. 2011. *Debt: The First 500 Years*. Brooklyn, New York: Melville House

Graves, William W. 2004. The Geography of Finance and Financial Services. *The Industrial Geographer* 2(1): 1.

Green, Stephen. 2000. Negotiating with the Future: The Culture of Modern Risk in Global Financial Markets. *Environment and Planning D: Society and Space.* 18: 77–89.

Haldane, Andrew G. 2011. *The Race to Zero*. Speech at the International Economic Association Sixteenth World Congress, Beijing. Retrieved from http://www.bankofengland.co.uk/publications/Documents/speeches/2011/speech509.pdf

Hall, Sarah. 2010. Geographies of Money and Finance I: Cultural Economy, Politics, and Place. *Progress in Human Geography* 35(2): 234–245.

Hart, Keith. 2001. Money in an Unequal World. *Anthropological Theory* 1(3): 307–330.

Hart, Keith. 2005. Notes Towards an Anthropology of Money. *Kritikos: an international and interdisciplinary journal of postmodern cultural sound, text, and image* 2, Retrieved from http://intertheory.org/hart.htm

Hart, Keith. 2007. Money is Always Personal and Impersonal. *Anthropology Today* 23(5): 12–16.

Hartsock, Nancy C.M. 1983. *Money, Sex, and Power: Toward a Feminist Historical Materialism*. New York: Longman.

Harvey, David. 1990. *The Condition of Post-Modernity*. London: Blackwell.

Hayek, Friedrich A. 1974. *Denationalisation of Money: The Argument Refined.* Auburn: Institute of Economic Affairs.

Heath, Eugene. 2010. 'Fairness in Financial Markets'. In Boatright, John R. (ed.). *Finance Ethics*. Hoboken, New Jersey: John Wiley & Sons, pp. 163–178.

Heeks, Richard. 2009. Real Money from Virtual Worlds. *Scientific American.* 302: 68–73.

Heidegger, Martin. 1927. *Being and Time* (trans. J. Stambaugh). Albany, NY: SUNY Press, 1996.

Heidegger, Martin. 1971. 'The Thing'. In *Poetry, Language, Thought* (trans. Albert Hofstadter) (pp. 161–184), New York: Harper & Row, 2001.

Heidegger, Martin. 1977. 'The Question Concerning Technology'. In *The Question Concerning Technology and Other Essays* (trans. William Lovitt) (pp. 3–35). New York: Harper & Row.

Himma, Kenneth Einar. 2009. Artificial Agency, Consciousness, and the Criteria for Moral Agency. *Ethics and Information Technology* 11(1): 19–29.

Hudson, Michael. 2004. The Archeology of Money: Debt versus Barter Theories of Money's Origins. In Wray, L. Randall (ed.). *Credit and State Theories of Money*. Cheltenham: Edward Elgar Publishing.

Hughes, Alex. 2005. Geographies of Exchange and Circulation: Alternative Trading Spaces. *Progress in Human Geography* 29(4): 496–504.

Hurlburt, George F., Miller, Keith W. & Voas, Jeffrey M. 2009. An ethical analysis of automation, risk, and the financial crises of 2008. *IT Professional* 11(1): 14–19.

Kirwan, James. 2004. Alternative Strategies in the UK Agro-Food System: Interrogating the Alterity of Farmers' Markets. *Sociologia Ruralis* 44(4): 395–415.

Knorr Cetina, Karin & Bruegger, Urs. 2002a. Global Microstructures: The Virtual Societies of Financial Markets. *American Journal of Sociology* 107(4): 905–950.

Knorr Cetina, Karin & Bruegger, Urs. 2002b. Inhabiting Technology: The Global Lifeform of Financial Markets. *Current Sociology* 50(3): 389–405.

Knorr Cetina, Karin. 2005a. 'From Pipes to Scopes: The Flow Architecture of Financial Markets'. In Barry, Andrew and Slater, Don (eds). *The Technological Economy*. London: Routledge.

Knorr Cetina, Karin 2005b. 'How are Global Markets Global? The Architecture of a Flow World'. In Knorr Cetina, Karin and Preda, Alex (eds). *The Sociology of Financial Markets*. Oxford: Oxford University Press.

Kolb, Robert W. 2010a. 'The Finance View of the World and its Ethical Implications'. In Boatright, John R. (ed.) *Finance ethics*. Hoboken, New Jersey: John Wiley & Sons (pp. 23–43).

Kolb, Robert W. 2010b. (ed.) *Lessons from the Financial Crisis*. Hoboken, New Jersey: John Wiley & Sons.

Laidlaw, James. 2000. A Free Gift Makes No Friends. *Journal of the Royal Anthropological Institute* (N.S.) 6(4): 617–634.

Lastowka, F. Gregory and Hunter, Dan. 2004. The Laws of the Virtual Worlds. *California Law Review* 92(1). Retrieved from http://scholarship.law.berkeley.edu/cgi/viewcontent.cgi?article=1345&context=californialawreview

Latour, Bruno. 1987. *Science in Action*. Cambridge, MA: Harvard University Press.

Lehtonen, Turo-Kimmo & Pyyhtinen, Olli. 2008. On Simmel's Conception of Philosophy. *Continental Philosophy Review* 41(3): 301–322.

Lenterman, R. 2013. Some HFT Myths Debunked. *The Financial Times*, March 1, 2013. Retrieved from http://www.ft.com/cms/s/0/7c7025ec-802e-11e2-aed5-00144feabdc0.html#axzz2TLdCbjFE

Leyshon, Andrew & Lee, Roger. 2003. 'Introduction: Alternative Economic Geographies'. In Leyshon, Andrew, Lee, Roger, & Williams, Colin C. (eds). 2003. *Alternative Economic Spaces*. London: Sage (pp. 1–26).

Leyshon, Andrew & Thrift, Nigel J. 1997. *Money/Space: Geographies of Monetary Transformation*. London/New York: Routledge.

Lin, Mingfeng, Viswanathan, Siva, and Prabhala, N.R. 2009. Judging Borrowers by the Company They Keep: Social Networks and Adverse Selection in Online Peer-to-Peer Lending. Retrieved from https://server1.tepper.cmu.edu/seminars/docs/viswanathan_paper.pdf

Lothian, James R. (2002). The internationalization of money and finance and the globalization of financial markets. *Journal of International Money and Finance* 21(6): 699–724.

Low, Will and Davenport, Eileen. 2006. Mainstreaming Fair Trade: Adoption, Assimilation, Appropriation. *Journal of Strategic Marketing* 14(4): 315–328.

MacKenzie, Donald and Millo, Yuval. 2003. Constructing a Market, Performing Theory: The Historical Sociology of a Financial Derivatives Exchange. *American Journal of Sociology* 109(1): 107–145.

MacKenzie, Donald. 2005. Opening the Black Box of Global Finance. *Review of International Political Economy* 12(4): 555–576.

MacKenzie, Donald. 2007. The Material Production of Virtuality: Innovation, Cultural Geography and Facticity in Derivatives Markets. *Economy and Society* 36(3): 355–376.

Martin, Felix. 2013. *Money: The Unauthorized Biography*. London: The Bodley Head.

Martin, Ron. 2011. The Local Geographies of the Financial Crisis: From the Housing Bubble to Economic Recession and Beyond. *Journal of Economic Geography* 11(4): 587–618.

Marx, Karl. 1844a. Economic and philosophic manuscripts of 1844. In *Economic and philosophic manuscripts of 1844 and the Communist Manifesto* (trans. Martin Milligan). Amherst, NY: Prometheus, 1988.

Marx, Karl. 1844b. 'On the Jewish Question'. *Deutscher-Fransösischer Jahrbücher.* Published as e-book *On the Jewish Question.* Aristeus Books, 2012. Also available at http://www.marxists.org/archive/marx/works/1844/jewish-question/

Marx, Karl. 1867. *Capital: A critique of political economy* Vol. I. (trans. Ben Fowkes). London: Penguin, 1976/1990.

Marx, Karl & Engels, Friedrich. 1846. The German ideology. In Marx, Karl & Engels, Friedrich, *Karl Marx and Frederick Engels Collected Works* Vol.5 New York: International Publishers/Moscow: Progress Publishers, 1976.

Maurer, Bill. 2006. The Anthropology of Money. *Annual Review of Anthropology* 35: 15–36.

Maurer, Bill, Nelms, Taylor C., and Lana Swartz. 2013. 'When perhaps the real problem is money itself!': The practical materiality of Bitcoin. *Social Semiotics* 23(2): 261–277.

Mauss, Marcel. 1925. *The Gift*. London: Routledge, 1990.

McLuhan, Marshall. 1964. *Understanding Media*. Abingdon/New York: Routledge, 2001.

McLuhan, Marshall & Fiore, Quentin. 1967. *The Medium is the Massage: An Inventory of Effects*. Harmondsworth: Penguin.Monnet, Jean. 1943. Jean Monnet's Thoughts on the Future. CVCE/Fondation Jean Monnet pour l'Europe. Retrieved from http://www.cvce.eu/content/publication/1997/10/13/b61a8924-57bf-4890-9e4b-73bf4d882549/publishable_en.pdf

Morgan, Edward Victor 1965. *A history of money* (Vol. 699). Harmondsworth: Penguin.

Muniesa, Fabian, Millo, Yuval, & Callon, Michel. 2007. An Introduction to market devices. *The Sociological Review* 55(s2): 1–12.

Muniesa, Fabian & Lenglet, Marc. 2013. Responsible Innovation in Finance: Directions and Implications. In Owen, Richard, Bessant, John, and Heintz, Maggy (eds). *Responsible Innovation: Managing the Responsible Emergence of Science and Innovation in Society*. Chichester: John Wiley & Sons (pp. 185–198).

Nakamoto, Satoshi. 2008. *Bitcoin: A Peer-to-Peer Electronic Cash System*. Retrieved from https://bitcoin.org/bitcoin.pdf

Nakamura, Lisa. 2009. Don't Hate the Player, Hate the Game: The Racialization of Labor in World of Warcraft. *Critical Studies in Media Communication* 26(2): 128–144.

Nelson, Anitra. 1999. *Marx's Concept of Money: The god of commodities*. London: Routledge.

North, Peter. 2005. Scaling Alternative Economic Practices? Some Lessons from Alternative Currencies. *Transactions of the Institute of British Geographers* NS 30(2): 221–233.

Pearson, Ruth. 2003. Argentina's Barter Network: New Currency for New Times? *Bulletin of Latin American Research* 22(2): 214–230.

Pinch, Trevor and Swedberg, Richard (eds). 2008. *Living in a Material World: Economic Sociology Meets Science and Technology Studies*. Cambridge, MA/London: MIT Press.

Plessner, Helmuth. 1928. *Die Stufen des Organischen und der Mensch*, Gesammelte Schriften, vol. 4. Frankfurt am Main: Suhrkamp, 1981.

Prior, Francesc and Argandoña, Antonio. 2009. Best Practices in Credit Accessibility and Corporate Social Responsibility in Financial Institutions. *Journal of Business Ethics* 87: 251–265.

Pryke, Michael & Allen, John. 2000. Monetized Time-Space: Derivatives – Money's 'New Imaginary'? *Economy and Society* 29(2): 264–284.

Raynolds, Laura T. 2000. Re-Embedding Global Architecture: The International Organic and Fair Trade Movements. *Agriculture and Human Values* 17(3): 297–309.

Reynolds, John N. 2011. *Ethics in Investment Banking*. Basingstoke/New York: Palgrave Macmillan.

Roberts, Keith. 2011. *The Origins of Business, Money and Markets*. Columbia University Press.

Robertson, Roland. 1995. 'Glocalization: Time-Space and Homogeneity-Heterogeneity'. In Featherstone, Mike, Lash, Scott, & Robertson, Roland (eds.) *Global Modernities*, pp. 25–44. London: Sage.

Robinson, William I. 2007. 'Theories of Globalization'. In Ritzer, George (ed.). *The Blackwell Companion to Globalization,* Malden, MA/Oxford/Carlton, Victoria: Blackwell (pp. 125–143).

Rose, Gregory M. & Orr, Linda M. 2007. Measuring and Exploring Symbolic Money Meanings. *Psychology & Marketing* 24(9): 743–761.

Rosenau, James N. 2003. *Distant Proximities: Dynamics beyond Globalization*. Princeton: Princeton University Press.

Ross, Alice K., Mathiason, Nick, & Fitzgibbon, Will. 2012. Robot Wars: How high frequency trading changed global markets. *The Bureau of Investigative Journalism*. Retrieved from http://www.thebureauinvestigates.com/2012/09/16/robot-wars-how-high-frequency-trading-changed-global-markets/

Sassen, Saskia. 1991. *The Global City: New York, London, Tokyo*. Princeton: Princeton University Press.

Scholte, Jan Aart. 2002. 'What is Globalization? The Definitional Issue – Again'. The University of Warwick Centre for the Study of Globalisation and Regionalisation Working Paper No. 109/02, December 2002. Retrieved from http://wrap.warwick.ac.uk/2010/1/WRAP_Scholte_wp10902.pdf

Searle, John R. 1995. *The Construction of Social Reality*. New York: The Free Press.

Searle, John R. 2005. What is an institution? *Journal of Institutional Economics* 1(01): 1–22.

Searle, John R. 2006. Social Ontology. *Anthropological Theory* 6(1): 12–29.

Seyfang, Gill. 2000. The Euro, the Pound and the Shell in our Pockets: Rationales for Complementary Currencies in a Global Economy. *New Political Economy* 5(2): 227–246.

Seyfang, Gill. 2001. Money That Makes a Change: Community Currencies, North and South. *Gender and Development* 9(1): 60–69.

Seyfang, Gill. 2002. Tackling Social Exclusion With Community Currencies: Learning From LETS to Time Banks. *International Journal of Community Currency Research* 6. Retrieved from https://ijccr.files.wordpress.com/2012/05/ijccr-vol-6-2002-3-seyfang.pdf_

Sherratt, Susan and Sherratt, Andrew. 1993. The Growth of the Mediterranean Economy in the Early First Millennium BC. *World Archaeology* 24(3): 361–378.

Simmel, Georg. 1907. *The Philosophy of Money* (3rd edition), ed. Frisby, David (trans. Bottomore, Tom & Frisby, David). London & New York: Routledge, 2004.
Simmel, Georg. 1997. 'The Crisis of Culture'. In Frisby, David & Featherstone, Mike (eds). *Simmel on Culture*, London: Sage.
Singh, Supriya. 2013. *Globalization and Money: A Global South Perspective*. Lanham, Maryland/Plymouth: Rowman & Littlefield.
Stahl, Bernd Carsten. 2004. Information, Ethics, and Computers: The Problem of Autonomous Moral Agents. *Minds and Machines* 14(1): 67–83.
Stulz, René M. 2005. The Limits of Financial Globalization. *The Journal of Finance* 60(4): 1595–1638.
Tasch, Woody. 2008. *Inquiries into the Nature of Slow Money: Investing as if Food, Farms, and Fertility Mattered*. White River Junction, VT: Chelsea Green.
Venkatesan, Soumhya. 2011. The social life of a 'free' gift. *American Ethnologist* 38(1): 47–57.
Virilio, Paul. 1977. *Speed and Politics*. (trans. Semiotext(e) & Polizzotti, Mark) Los Angeles, CA: Semiotext(e), 2006.
Von Schomberg, René. 2013. 'A Vision of Responsible Research and Innovation'. In Owen, Richard, Bessant, John, & Heintz, Maggy (eds.). *Responsible Innovation: Managing the Responsible Emergence of Science and Innovation in Society*. London: John Wiley & Sons (pp. 51–74).
Wallach, Wendell. and Allen, Colin. 2009. *Moral Machines: Teaching Robots Right from Wrong*. Oxford/New York: Oxford University Press.
Weber, Max. 1905. *The Protestant Ethic and the Spirit of Capitalism* (trans. Talcott Parsons). London and New York: Routledge, 1992.
Zaloom, Caitlin. 2003. Ambiguous Numbers: Trading Technologies and Interpretation in Financial Markets. *American Ethnologist* 30(2): 258–272.
Zaloom, Caitlin. 2006. *Out of the Pits: Traders and Technology from Chicago to London*. Chicago: University of Chicago Press.
Zelizer, Viviana A. 1989. The Social Meaning of Money: 'Special Monies'. *The American Journal of Sociology* 95(2): 342–377.
Zelizer, Viviana A. 1997. *The Social Meaning of Money*. Princeton, NJ: Princeton University Press.
Zhang, Lin and Fung, Anthony Y.H. 2013. Working as Playing? Consumer Labor, Guild and the Secondary Industry of Online Gaming in China. *New Media & Society* 16(1): 38–54.

# Index

**Bold** page numbers indicate figures.

accountability 31
accounting
  as financial technology 20
  recording devices 28–9
activities and humans, distancing between 5
actor-networks 134, 135, 162
agencement 138
agricultural revolution 18–21
algorithms 108–9, 118, 140
alienation 47–8, 58–9, 70–3, 79–80, 145
Allen, J. 81, 131–2, 143, 155–6
alternative technologies and practices
  barter 159
  Bitcoin 163–5
  computer games, money in 165–8
  euro 170–2
  Fair Trade network 157, 162
  future for 175–6
  gifts 174–5
  Local Exchange Trading Systems (LETS) 159–61, 161
  local food production and consumption 157–8
  microcredits 169–70
  money and finance, forms of 158–75
  need for 151
  non-monetised social economy 160
  peer-to-peer lending 169–70
  and resistance 152–6
  Slow Food movement 161
  Slow Money 172–4
  socio-technical systems 162
  time currencies/banks 160–1
  trade and production practices 157–8
ancient world 18–21, 26–8
anthropology of finance 9–10, 134, 136–7, 143–4
arbitrage 138–9
Arendt, Hannah 70–3
Argandoña, A. 170
Aristotle 113
artificial intelligence in global finance 108–10
Ashta, A. 169
Assadi, D. 169
assemblages 138
automation
  and control 118–19
  in global finance 108–10
  and market uncertainty 111

banks
  access to services 124–5
  development of 24–5
  and financial technologies 25–6
  time 160–1
Barry, A. 135–6
barter 18, 21, 159
Beunza, D. 116, 138–9
Bitcoin 104
  as alternative practice 163–5
  independence from nation states 105–6
  information ontology 96
  materialism of 140–1
  object ontology 94
  as political 163–5
  remediation 56–7
  trust between peers 105–6
blending 75
Boatright, J.R. 7
bookkeeping
  as financial technology 20
  technologies and distancing 30–2
Borgman, A. 48
Brey, P. 75
Brignall, T. 168
Bronk, C. 167
Bruegger, U. 78–9, 114, 127–8, 138

business ethics 6–7

calculating machines 29–32
calculation
and judgement 141–2
social and material aspects of 144
Callon, M. 134, 135, 144
*Capital* (Marx) 58, 155
capitalism 152–3
Castells, M. 73–5, 127
Castronova, E. 166–7, 167n3
categorization problems raised by money 92–5
change, financial-social, in humans and technologies 145
*see also* alternative technologies and practices
Clark, G.L. 9
cockpits, trade 139–40
collective intentionality 98–9
community, new forms of 154
compression by electronic technologies 77–8
computer games, money in 165–8
*Construction of Social Reality, The* (Searle) 98
control and responsibility 117–19, 120
credit 24–5
culture
implications of financial globalization 80–2
of specific practices 143–4

daily life, impact of global finance on 132
Davies, W. 9
Davis, M. 111, 112
de-socialization of work spaces 114
debt 24–5
deep-relational view of money 99–101, 103–4
dematerialization 53–4
digital gap 124–5
digital media. *see* information and communication technologies (ICTs)
digital objects 92–5
disembedding 80
distance/distancing
agricultural revolution 20–1
all-purpose tools 49
calculating machines 30–2
epistemic condition 113–17
and financial technologies 3–4, 5–6
food 155–6
globalization 45
between humans and their activities 5
and information and communication technologies (ICTs) 4–5
between medium and message 42
and money 22–3
money as creating 41
money as medium 35–7
money as pure symbol and function 37–8
nature, distancing from 41
objective/subjective worlds in modernity 50–3
as part of what we do 145
persons and goods 45
and responsibility 113–20
responsibility for financial transactions 30–2
un-distancing through technology 156–75
*see also* alternative technologies and practices; resistance to distancing
distance-proximity binary 126
division of labour 84–5
Dobson, J. 7
Dodd, N. 160
double orientation in global trading 127–8
Dreyfus, H.L. 48, 76

earth, alienation from 70–3
*Economic and Philosophic Manuscripts Of 1944* (Marx) 47, 58, 59, 79
economic citizenship 161
economics 135, 144
Egypt 19
electronic money
and anonymity 44
collective intentionality 98–9
computer games, money in 165–8
as detached from substance 43
distancing effects 53
ethical issues 90
and globalization 44–5
information ontology 96

meaning of money 89
metaphysics of 91–101
as not fully omnipotent 44
object ontology 94
possession of as end in itself 43
remediation 56–7
and responsibility 77
and Simmel 42–6
*see also* Bitcoin
embodiment 128, 138
empowerment, promise of through technology 153–4
epistemic condition 113–17
Ess, C. 96
essentialism 55–6
ethics
electronic monies 90
human and subjective aspect of finance 143–4
insights from 6–7
and technology 111–12
euro 170–2
experience
directness and remoteness of 23
in the global village 75–82

Fair Trade network 157, 162
farmers markets 158
Ferguson, N. 23
financial crisis 2008 132
financial products, epistemic problems with 116–17
financial technologies
as changing space 131–2
and distance 3–4
and distancing 5–6
division of labour 84–5
future for 175–6
lack of attention given to 3
material infrastructure of 134–41
meaning of 103–4
money as medium 82–7
and more integral form of life 85–6
non-neutrality of 136
objective/subjective worlds in modernity, distancing between 52–3
personalization and socialization of 142–3
questioning the nature and meaning of 91
and responsibility 77
*see also* alternative technologies and practices
Fine, B. 146–7
Flash Crash of May 2010 109
Fletcher, J.H. 163
Floridi, L. 95–6, 97, 99, 134
food
resistance v. reform 155–6
Slow Food movement 161
speculation on 133
Foodwatch 133

Galbraith, J.K. 24
games, online, money in 165–8
geographical disembedding thesis 75
geography of finance 9–10
refinement of 129–34
*see also* globalization; local, the
Giddens, A. 66
gifts 174–5
Gilbert, E. 130
global and local, entanglement of 126
*see also* globalization; local, the
Global Barter Network (Argentina) 159
global responsibility 75–82
global village 75
globalization
concept of 64–7
and distance 45
electronic monies 44–5
financial 78–82
financial, implications of for culture 80–2
history of 26–8
and ICTs 67–70
less moral distance in 75–6
limits and barriers to 124–5
and money 38–9
place, continued importance of 54
as spatial and social process 64–7
work spaces 114–15
*see also* local, the
glocalization 126
*see also* globalization; local, the
Greece, ancient 26–7
Green, S. 143

Haldane, A. 109
Hart, K. 145–6, 147–8, 153, 154
Harvey, D. 66
Hayek, F.A. 164
Heath, E. 7
Heidegger, M. 47, 66, 77, 91
high-frequency trading (HFT) 108–9, 117
history of financial technologies
  agricultural revolution 18–21
  ancient world 18–21
  banks 24–6
  barter system 18
  broader significance of 18
  calculating machines 29–30
  debt and credit 24–5
  globalization, history of 26–8
  money 21–6
  recording devices 28–9
  writing 20
horizontal responsibility for financial transactions 31–2
*Human Condition, The* (Arendt) 70–2
human essence 55–6
human geography 9
humanization of financial technologies 142–3
humans and their activities, distancing between 5
Hurlburt, G.F. 111

impersonal side to global finance 145–8
inequality, technologies as alleviating 153–5
information, philosophy of 95–6
information age 73
information and communication technologies (ICTs)
  as changing social relations 68
  concerns over role in finance 2
  content of as important 45–6
  and distancing 5–6
  distancing effect of 4–5
  empowerment, promise of through 153–4
  globalization 67–70
  objective realm created by 42
  as purest example of the tool 43
  and Simmel 42–6
  *see also* electronic monies
information ontology of money 95–7, 99–100, 102, 103
*Information Society, The* (Ess) 96
internalization of calculation and accountability 31–2
internet. *see* information and communication technologies (ICTs)
intersubjectivity in global trading 127–8

Kirwan, J. 157, 158
Kiva 169
Knorr Cetina, K. 78–9, 80, 114, 127–8, 129, 136–7, 138
knowledge
  directness and remoteness of 23
  Fair Trade network 162
  and responsibility 113–17, 119–20
Kolb, R.W. 7

labour
  alienated 47–8, 58–9
  division of labour 84–5
language 98, 103
Lapavitsas, C. 146–7
Latour, B. 134, 139
*Laws of the Markets, The* (Callon) 135, 144
Lee, R. 156, 162
leisure-work 84–5
Leyshon, A. 135, 156, 162
Linden dollars 166
local, the
  daily life, impact of global finance on 132
  embodiment 128
  financial crisis 2008 132
  food, speculation on 133
  food production and consumption 157–8
  and global, entanglement of 126
  in global trading 127–
  physical presence of firms 129
  place, continued importance of 54, 127, 128–9
  resistance to distancing 156
  *see also* globalization
Local Exchange Trading Systems (LETS) 159–60, 161

MacKenzie, D. 9, 144
market, the 80
Martin, F. 19, 22, 132
Marx, K. 47, 57–9, 79–80, 130, 152, 155
material infrastructure
  Bitcoin 140–1
  continued importance of 129
  of financial technologies 134–41
  as shaping financial world 137–8
Maurer, B. 141, 164–5
McGoey, L. 9
McLuhan, M. 42, 49, 53, 67–9, 75–8, 82–7
meaning-giving in financial practices 143
media theory, remediation in 56–7
medium is the message 5, 42, 49, 53, 67–8, 70
Mesopotamia 19, 26
metaphysics of money
  deep-relational view of money 99–101, 103–4
  electronic money 91–101
  information ontology 95–7
  money and institutions 104–6
  money as a technology 101–4
  need for 89–90
  object ontology **92**, 92–5
  questioning the nature and meaning of money 91
  social ontology 97–9
  variety of meanings for money 91
microcredits 169–70
Millo, Y. 144
modernity
  alienation 47–8
  objective/subjective worlds, distancing between 50–3
  social relations in 39–42
money
  before 18
  as absorber of all values 58
  alienated character of 58
  as a commodity 57–9
  in computer games 165–8
  as creating distance 41
  debt 24–5
  deep-relational view of 99–101, 103–4
  as the final end 38
  as freedom tool 22
  and globalization 38–9
  history of 21–6
  individuals and possession, distance between 40–1
  and institutions 104–6
  as instrument of rule 22
  meaning of 89
  as medium 35–7, 82–7
  and moral distance 22–3
  and nation states 104–6
  nature, distancing from 41
  objective/subjective worlds in modernity, distancing between 50–3
  and power relations 59
  as pure information 84, 86–7
  as pure symbol and function 37–8
  as reducing direct relations 23
  remediation 56–7
  situated in time and space 130
  and social organization 22
  social organization and changes in form of 83–4
  and social relations in modernity 39–42
  as symbol of abstraction 41
  as a technology 101–4
  as a tool 38
  utopian character of 160
  *see also* electronic monies; metaphysics of money; ontologies of money
*Money: The unauthorized biography* (Martin) 19
moral distance/distancing
  agricultural revolution 20–1
  epistemic condition 113–17
  and information and communication technologies (ICTs) 4–5, 5–6
  and money 22–3
  and responsibility 113–20
Morgan, E.V. 21, 22, 24, 25
multidisciplinary approach 10
Muniesa, F. 138

Nakamoto, S. 165
Nakamura, L. 167
nation states 104–6, 124, 163
nature, distancing from 41
network society 73, 134–6, 162
*Nicomachean Ethics* (Aristotle) 113

non-monetised social economy 160
North, P. 161
numbers 93–4, 141–2

object ontology of money **92,** 92–5, 99–100, 102–3
object-subject relation 100
objects
  impersonal character of 40
  value of as objectified 35
'On the Jewish Question' (Marx) 57, 79–80
ontologies of money
  beyond 99–101
  information ontology 95–7, 99–100, 102, 103
  object ontology **92,** 92–5, 99–100, 102–3
  social ontology 97–9, 102, 103

parallel universes 74, 127
peer-to-peer lending 169–70
personal side to global finance 145–8
personalization of financial technologies 142–3
philosophy of information 95–6
philosophy of technology
  empirically oriented approach of 7–8
  Simmel's contribution to 47–50
physical presence of firms 129
physicality in global finance 138
Pinch, T. 135
place, continued importance of 54, 127, 128–9, 129, 173
play and work, end of separation of 167–8
political association, new forms of 154
power
  capitalism 152–3
  collective intentionality 98–9
  money and 59
  resistance to structures of 152
Prior, F. 170
professional teams, knowledge problems with 115
Pryke, M. 81, 131–2, 143

'race to zero, The' (Haldane) (speech) 109
Raynolds, L.T. 157
recording devices 28–9
remediation 56–7
resistance to distancing
  and alternatives 152–6
  local 156
  need for 151
  *see also* alternative technologies and practices
responsibility
  and control 117–19, 120
  epistemic condition 113–17
  and financial technologies 77
  for financial transactions 30–2
  food, speculation on 133
  in the global village 75–82
  human and subjective aspect of finance 143–4
  improving conditions for exercising 119–20
  and knowledge 119–20
  local impact of global finance 133
  material infrastructure of financial technologies 136
  microcredits 169–70
  and moral distance 113–20
  ways to exercise 112
Reynolds, J.N. 7
risk cultures 143–4
Roberts, K. 18, 19, 26, 27
Robertson, R. 126
Rome, ancient 27
Rosenau, J.N. 126

Sassen, S. 67, 128–9
Scholte, J.A. 65, 66, 67, 125–6
science and technology studies (STS) 8–10, 134, 143–4
screen-based trading 138
Searle, J. 98–9, 103
Second Life 166
self-surveillance 31–2
Seyfang, G. 158–60, 161
Sherratt, A, 27–8
Sherratt, S. 27–8
Simmel, Georg 130
  dematerialization 53–4
  deterministic, analysis as 54–5
  distance, money as creating 41
  distance between medium and message 42
  electronic monies 42–6

financial globalization 81
human essence 55–6
individuals and possession, distance between 40–1
and Marx 57–9
and McLuhan 82–7
money and social relations in modernity 39–42
money as a tool 38
money as medium 35–7
money as pure symbol and function 37–8
objective/subjective worlds in modernity, distancing between 50–3
philosophy of technology, contribution to 47–50
place, continued importance of 54
problems with analysis by 50–60
Slater, D. 135–6
slave labour 26
Slow Food movement 161
Slow Money 172–4
social media 169
social ontology of money 97–9, 102, 103
social organization
calculating machines 29–30
and changes in form of money 83–4
ICTs as changing 68
in modernity 39–42
and money 22
money as social institution 37
new, in the ancient world 19–20
and recording devices 28–9
social space
embodiment, role of in 128
as global and local 126, 127–8
as territorial and supraterritorial 125
social studies of finance 8–10, 136–41
socialization of financial technologies 142–3
society, implications of financial globalization for 80–2
space, finance and its technologies as changing 131–2
space of flows 73–5, 80
spatial dimension of globalization 64–7
standard means of exchange 21–2
Stark, D. 116, 138–9
states, nation 104–6, 124, 163
subject-object relation 100
subjectivity 31–2
surveillance, self- 31–2
Swedberg, R. 135

Tasch, W. 172–4
technology
alienation and exploitation 47–8
as alleviating inequality 153–5
compression by electronic technologies 77–8
concerns over role in finance 2
and ethics of finance 111–12
medium and content, importance of 49–50
money as a 101–4
as purest example of the tool 43
take over by 55
un-distancing through 156–75
unintended consequences 51, 67–9
wider impact of 48–9
*see also* electronic monies; financial technologies
Thrift, N.J. 135
time currencies/banks 160–1
tools
all-purpose, distancing effects of 49
money as 38
people as all-purpose tools 46
technology as purest example of 43
unintended effects of 48–9
trade and financial technologies 27–8
trade cockpits 139–40
trading pits 138
transdisciplinary approach 10

*Understanding Media* (McLuhan) 68, 75, 82
utopian character of money 160

vertical responsibility for financial transactions 30–1
virtual objects 92–5
virtual worlds, money in 165–8
vulnerability of global financial system 129

Weber, M. 47, 130
work and play, end of separation of 167–8

work-leisure 84–5
work spaces, globalization of 114–15
world alienation 70–3
World of Warcraft 166, 168
writing
  as financial memory tool 27
  as financial technology 20

Zaloom, C. 113, 138, 141–2
Zelizer, V.A. 9, 146